MOLYMANIA
THE GREAT MOLYBDENUM BULL MARKET

By Michael H. Caldwel

Creative Classics Inc.

Kelowna, BC

Canada

ISBN: 978 – 0 – 9784620 – 0 – 0

Caldwell, Michael H., 1946 –

MolyMania – the Great Molybdenum Bull Market

Printed by: Friesen Printing Corporation

Introduction by: Marc Davis

Research: Peter Mitham, Toby Osborne, Stan Sauerwein, Tony Wanless

PRINTED IN CANADA

TABLE OF CONTENTS

INTRODUCTION

Molybdenum!

It's fast becoming a fashionable word that many of us couldn't even pronounce not that long ago. That's not surprising since this historically unglamorous, little-known industrial base metal was barely trading above a paltry $2/lb during much of 2002. This made molybdenum exploration a pretty unattractive proposition for most mineral exploration companies. And it was equally shunned by most investors.

There again, 2002 was also the year that most metal commodity prices were in a slump, including gold. Politically, it was also a troubled time in the aftermath of 9/11. Remember President George W. Bush's emphatic warnings to the world that the U.S. may be forced to mobilize for war with Iraq? All because of the need to destroy Saddam Hussein's alleged "weapons of mass destruction" and to combat terrorism.

How quickly things change. But on a brighter note, it's sometimes for the best. That certainly proved to be the case over the past five years for the base and precious metals markets and the mining industry, alike. And arguably the unlikely star performer among the metals sector has been the one with the tongue-twisting name that has since become better known by its more user-friendly abbreviation, "moly."

Its meteoric rise to a peak of around the $45/lb mark in mid 2005 proved as dramatic as the global economic boom that fuelled this remarkable ascendancy. In particular, a resurgent oil and gas industry and strong demand from China proved to be the main catalysts behind this reversal of fortunes. Since then burgeoning demand for molybdenum has continued to outstrip supplies. So much so that as of the end of 2007 prices still remain incredibly buoyant in the 30/lb-35/lb range.

Indeed, that's particularly good news for most Canadian "moly" exploration companies, many of whom have the benefit of focusing their efforts and expertise on their own back yards. That's because Canada is a vast, geologically fertile hunting ground for new molybdenum discoveries. Similarly, a sustained 1,500-2,000% spike in molybdenum prices since 2002 has also proved to a major shot in the arm for companies that are advancing existing in-development deposits.

These exciting developments promise to help ensure that Canada continues to be a world leader in molybdenum production with annual output at around 20 million pounds and climbing.

Following are eye-opening excerpts from an article entitled "Obscurities Rise to Prominence" that was published by the online investment newsletter, PinnacleDigest.com, on October 7, 2007:

"We have discussed molybdenum before but never with this urgency and powerful documentation that will create only one thing: Demand.

Just for a minute don't think about the market or demand for any particular commodity; think about what we need to continue our way of living. Think about how dependant we actually are on essentials such as heating and cooling our homes or driving our cars. Beneath our roads, homes, farms, deserts, malls and stadiums lies a multi-million mile network of aging pipelines which supply our energy needs.

These pipelines are our energy transportation system, which we rely on to fuel our cars and power our computers. The corroding of these pipelines is due to hydrogen sulphide, carbon dioxide and common oxygen. How serious is the problem and how much money and demand will be involved? Corrosion costs the U.S. economy over $276 billion a year. This number has no where to go but up.

A very high percentage of energy providing vehicles have aged beyond their original design life span. We are talking about nuclear power plants and the U.S. pipeline network. Over half of the United States' 2.4 million miles of oil and gas pipelines were built in the 1950s and 1960s. These pipelines have deteriorated. Corrosion is a leading cause of pipeline spills or failures. The corrosion leads to leaks and ruptures constantly.

We all remember last August when BP had to shut roughly 8% of US oil production. "Unexpectedly severe corrosion" in their Alaska pipeline was stated. This was the first ever shutdown in America's largest oil fields. Immediately following the corroded pipeline rupture, the industry introduced legislation in an attempt to prevent a recurrence of the disaster.

The Pipeline, Inspection, Protection and Enforcement and Safety Act was signed into law. There is much information which can be uncovered in regard to this law, but its purpose and goals have not been realized in many respects. The pipeline industry took a look at the aging, out-dated infrastructure and aimed the legislation towards a low cost solution -- the detection of corrosion and bit by bit pipeline replacement. They conveniently forgot to deal with the issues which led to the initial problem.

During the 1960s, a massive build-up phase occurred. During this time the US pipeline infrastructure relied on carbon and low-alloy steels for natural gas and petroleum transportation. As oil fields have aged, so have the pipelines that transport the oil. The risk of pipeline corrosion has never been higher. If the United States is encountering this problem, what are the chances countries such as Russia, China and all of Europe will be encountering the same one¿ More than probable.

Molybdenum was initially used to harden steel and increase weld ability while reducing the carbon content previously utilized. This metal has played a critical role in the development of oil and gas projects in the arctic and sub-sea environments.

Last December, in a worldwide pipeline construction survey compiled by Rita Tubbs some amazing facts were shown. She stated that, "81,593 miles of new and planned oil and gas pipeline was under construction and planned." She also stated that North American pipeline construction plans nearly doubled to 28,314 miles.

Tubbs also cited that, "By 2008, contractors expect to see a workload that has not been seen in Canada for nearly three decades." This can only be referred to one thing: the oil sands of Alberta. Molybdenum will have to play a large role in moving oil sands by pipeline.

China currently has over 24,000 miles of existing oil and gas pipelines. They are currently adding 15,000 miles to it, aimed to be completed by 2010. Upon this completion, the country has reported that it hopes to extend its mileage by nearly 63%. In 2005, China completed a 625-mile pipeline from Kazakhstan, set to import 140 million barrels of oil annually.

The world's largest energy consumer by far is the United States. By 2025, the EIA has reported that the US will need 47% more oil and 57% more natural gas. The transportation and distribution lines needed to sustain this growth equates to a 30% increase in current pipelines, resulting in roughly 600,000 miles of new pipeline. We are willing to bet the United States will do and pay anything to achieve this.

Each pipeline mile could contain between 600 and 1000 pounds of molybdenum. The United States has over 2.4 million miles in pipeline. Roughly one half of their pipelines could call for replacements. That means that updating America's outdated pipelines may require more than 300 million and up to a possible one billion pounds of molybdenum.

These are speculative statements but the facts justify that as these pipelines age 30 or 40 years, they will have to be replaced. Remember that chemical changes in the material passing through the pipelines accelerate erosion. The extraction from artic and sub sea levels add pressure to the pipes, as well as lengthening them, which creates the need for more molybdenum. As China brings on new projects, Russian and Europe share in our aging pipeline problem in North America.

The demand for molybdenum is something we believe will only increase. It even has environmental benefits in that it can be used to make pipelines and other metallic structures that weigh less and are more durable and corrosion-resistant than is presently the norm.

Currently molybdenum is roughly worth $30-35/lb and will be rising higher over the coming ten years. We believe that the molybdenum market has not peaked and that demand and absolute necessity of this metal will drive it into the public eye, affecting its value greatly.

There will be a select group of junior mining companies who will capitalize perfectly on the demand market for molybdenum which we are just entering."

Those companies that PinnacleDigest.com alludes to are the ones that have all the right ingredients for success: shrewd management, impressive geological expertise and technological know-how, ready access to generous sums of investment capital, and a liberal sprinkling of good luck.

That's the key to finding and carefully developing the next crop of economically viable "company maker" molybdenum deposits. The odds in favour of such successes may still be daunting and not for the feint of heart. However, patient investors with steady nerves stand to gain immeasurably from backing the best candidates.

Meanwhile, there's no shortage of intrepid, visionary companies profiled in this book that are taking investors on an exciting and profitable journey in their quest to prove up Canada's future molybdenum mines. In the process, some investors are likely poised to hit the ball right out of the park for a "home run" win. Will you, the reader, be one of them? Read on. And you decide.

Marc Davis, Managing Editor, SmallCapMedia.com

ADANAC MOLYBDENUM CORPORATION

Ruby Creek could be history in the making, in the eyes of Adanac Molybdenum Corporation's Executive Chairman, Larry Reaugh. His junior mineral-exploration company is assigned with the bold mission statement of turning their British Columbian moly property into "the first new large-scale, open-pit, primary molybdenum producer in 25 years."

"We're definitely in the lead for that," says Reaugh, positively. "Of course, we are focusing on the phase one, five-year pit which is high-grade 0.084% Mo—and, you know, that puts us up above some of the producers out there right now. So, Ruby Creek is going into production; starting construction in 2007."

Adanac, which was founded in 1992 and is based in White Rock, British Columbia, just south of Vancouver, acquired a 100% interest in Ruby Creek in 2000. Adanac's founder, Larry Reaugh has amassed an impressive 44 years in the mining industry, with the past 26 years directly involved in mineral exploration with junior resource companies. As the Executive Chairman of Adanac Molybdenum Corporation since October 2006, Reaugh was responsible for acquiring Ruby Creek, located close to Atlin, BC. He also instigated the business and development plan, and assembled an expert team to transition the project into a proposed world-class molybdenum property.

Initially named Stirrup Creek Gold Ltd., in October 2002 Reaugh's company received approval from its shareholders to change its name to. Two years later, the name was officially changed to Adanac Moly Corp. to reflect the company's primary exploration project, the Ruby Creek molybdenum deposit. The common shares of the company were first listed publicly on the TSX Venture Exchange in July 1995, and in May 2007 the company de-listed from TSX-V and listed their shares on TSX, under the same symbol: AUA.

Reaugh is conscious of the potential that is present at Ruby, and is determined to make something of it. As a man who is credited with several producing mines, having raised more than $200 million for the exploration and development of mining properties, he would recognize a contender for the next, big, open pit, moly mine if he saw one.

Reaugh says, “I’ve been focused on moly with a few other companies; Verdstone which is now GoldRea, and one called MolyCor. Adanac was primarily focused on gold here in British Columbia, where the company drilled out the high-grade gold resources at Watson Bar, and also in Nevada. The reality is, I was looking for tungsten properties and we went up and staked a couple of tungsten properties in B.C.; ended up picking up the Adanac pit (Ruby Creek was actually staked by an unrelated company called Adanac Mining and Exploration Co. Ltd). We didn’t even know it was open. So, we got it for minimal costs and then we’ve developed it all since then.”

Prior to becoming Executive Chairman in 2006, Reaugh held the position of President and Chief Executive Officer of Adanac since he founded the company in 1992. Yet, the B.C.-born, 63-year-old’s experience in mining dates back to his teenage years. “Soon after I graduated from high school I went to work for Bethlehem Copper (now part of Teck Cominco), which was an open pit mining operation, out of Ashcroft, in the Kamloops Mining Division.”

Starting as a warehouseman and later becoming a pit sorter and grade estimator, Reaugh eventually moved into survey and exploration work at Bethlehem. He went on to earn the position of Chief Surveyor for the company and then, from 1969 to 1973, Safety Engineer. Eventually Reaugh moved on to start his own company, L&L Drilling and Exploration.

Reaugh’s experience as President and CEO dates back to 1980, holding the esteemed positions at various mining companies, including Goldrea (from 1983 until 2006), MolyCor, Rocher Deboule, Midland Gold Corporation and Rea Gold. “So, I never actually studied. I’m a hands-on person. I’ve worked in every aspect of exploration, development and operation. And, I’ve been in the public arena for the past 27 years or so, running companies.” Over those years, Larry Reaugh has witnessed the turbulent ups and downs of the business, although now Adanac seems to have fallen on good times. “Yes, from 1997 up until about early 2003, mining just stagnated. I mean, it was difficult to raise money; it was difficult to get any interest in that area.”

Nevertheless, the Executive Chairman of Adanac didn’t allow this stagnant period to sap his spirit. “My major drive is to find, purchase and explore properties of merit that have the potential to go on into production,” he enthuses. “In my prior companies, we’ve had three discoveries that have gone into production. And I’ve had at least four or five properties that I’ve owned that have also gone into production, but we sold those properties.

Recognizing properties with potential is really key. And, you have to have talented people you're involved with."

The list of talented people at Adanac Molybdenum includes current President and CEO, Michael MacLeod. He has been an active Mining Engineer for 25 years and is a member of the Association of Professional Engineers and Geoscientists of B.C. (APEGBC). Furthermore, MacLeod has obtained a B.Sc. Eng. (Mining), an M. Eng. (Mining) and an MBA. MacLeod has been responsible for major capital projects, feasibility assessments and marketing and environmental studies.

MacLeod will be responsible for all project development and operational activities, and for assembling and leading the team of professionals that will advance the Ruby Creek moly project—a property recognized for its vast potential to deliver on Adanac's dream of opening a new, major, primary moly mine.

Identifying the moly trend in late 2003, Adanac implemented its plan to put its Ruby Creek molybdenum deposit into production by 2008. Thus far, the company has an eighteen month lead over any similar projects, and they expect to enjoy the highest moly prices, resulting in a rapid payback on their investment.

Larry Reaugh was well aware of the importance of getting a head start over the competition; in fact, he notes, "I've probably been one of the longest bulls on moly. We started back in the early 1990s, and nobody else really got going until the end of 2003. So, you know, we've been following that market and exploring in that market; it gave us an edge. We recognized the trend in 2003 and started our scoping studies immediately."

These studies indicate that Ruby Creek will be a 20,000 tonnes per day mine. Results published in January 2007 put Ruby Creek's mineable reserves at 143.7 million tonnes at 0.059% Mo, based on proven and provable ore in the amounts of 113.4 million tonnes grading 0.066% Mo, and 30.3 million tonnes at 0.034% Mo.

"296 million pounds of molybdenum. That's measured and indicated," states Reaugh.

The property sits at the headwaters of Ruby Creek, in the floor of an alpine cirque, at an elevation of 1,500 metres. Located about 22 kilometres northeast of Atlin, BC, and 124 kilometres southeast of Whitehorse, Yukon Territory, the Ruby Creek property lies in the extreme northwestern corner of B.C., and is accessible from Atlin by 39 kilometres of road.

Most of the logistics of the site are very favorable for mining, but electrical power has been a requirement that is in low supply in Atlin. A limited amount of hydro electrical power is available, while on site diesel electric power is generated to meet maximum requirements. However, surplus power is available from the Yukon Grid, which would only need the installation of 100 kilometres of transmission line—an option that was not available to prospective miners in the 1970s and 1980s.

"Ruby Creek is a property that was of interest to major companies who took it to feasibility; once in 1970 by Kerr Addison, and again in 1978 to 1980 by Placer Dome. But the price of moly always defeated them," says Adanac's Reaugh. "In 1970, it was $1.80 a pound and when Placer Dome did their feasibility, it was $6.00 a pound. Placer made the decision to go ahead with it, but then the price went below six. Luckily for us!" he laughs. "Or else, there'd be a big mine already going there right now."

Ruby Creek's illustrious history began with it being prospected and worked as long ago as 1898. Placer gold deposits have been the chief source of Atlin's economy. Even though the Ruby Creek molybdenum showing was first discovered in 1905 serious and systematic development didn't get underway until the late 1960s. The property was staked for Adanac Mining and Exploration Co. Ltd. (no relation to Adanac Molybdenum Corporation) in 1967. In 1968, an access road to the deposit and a camp were built, and surface exploration and initial diamond drilling were completed prior to shutting down for the winter season.

In the Spring of 1970, the Kerr Addison and Adanac Mining and Exploration finalized an agreement and a full scale feasibility program commenced, which including drilling and underground bulk sampling. A one hundred ton-a-day pilot milling operation together with related engineering studies was required as per the feasibility study. In April 1971, Chapman, Wood & Griswold Ltd. submitted a feasibility report on what was being called the Adanac Molybdenum project based on the Ruby Creek moly deposit. Chapman, Wood & Griswold reported an open pit estimate of 104,234,000 tons with an average grade of 0.16% MoS2. This is an historical estimate, not compliant with current National Instrument 43-101 standards.

Subsequently, in 1972, Climax Molybdenum Corp. of British Columbia optioned the property. Their option was dropped in 1975 and in 1978 an agreement was reached between Placer Development Limited and Adanac Mining and Exploration for further work and possible development of the Ruby Creek deposit.

The owners of the property from 1978 to 1982—Adanac, Kerr-Addison, Climax and Placer Developments—carried out extensive diamond drilling, comprising over 200 drill holes for a total of 31,962 metres. In addition, to facilitate geological mapping and sampling on the Adanac porphyry molybdenum occurrence, a 589-metre underground exploration drift was developed, including 246 metres of cross cut and 281 metres of raise. From 1982 on, no significant work on Ruby Creek was reported until the mineral claims on the deposit were dropped and it was open to be staked.

Seizing the opportunity, the company, then known as Adanac Gold Corp., announced in July, 2001 that it had acquired mineral claims in Boulder Creek, Surprise Lake, Atlin Mining Division. In October 2001, it was announced that Adanac had staked additional mineral claims in the area; at this time the claims completely covered the Ruby Creek molybdenum deposit. The National Instrument 43-101 report was completed on the property in April 2005—giving a 43-101 compliant 'measured and indicated' resource of 212.9 million tonnes grading 0.063% Mo at a 0.04% Mo cut-off. Exploration has continued onward to this day.

The preliminary feasibility was completed in a five month period , from April to September, 2005. The recommendation to proceed with a full bankable feasibility study was acted on immediately and feasibility was completed by April, 2006 with a recommendation to study the applicability of High Pressure Grinding Rolls (HPGR) technology as an alternate to conventional SAG (Semiautogenous) mill technology.

The HPGR process consists of a pair of counter rotating rolls; one is fixed and the other floating. The feed is introduced into the gap between the rolls for crushing. The grinding force applied to the crushing zone is controlled by a hydro-pneumatic spring on the floating roll. Due to indications of significant benefits for the project through more energy efficient crushing and grinding Adanac included HPGR in the updated bankable feasibility.

Moreover, during this period Adanac also established an encouraging relationship with the Taku River Tlingit First Nation, whose territory covers four million hectares in the northwest portion of B.C., including the 830-hectare mine development at Ruby Creek. The open pit mine would be located in the central portion of the Taku River Tlingit territory.

"We signed an agreement with the Taku River Tlingit over two years ago, and they've been part of the ongoing process for the socio-economic permitting. We have an excellent relationship with them," says Adanac's Larry Reaugh. With this valuable support, Adanac proceeded with its 2007 drill program at Ruby Creek. Fourteen holes, totaling 7000 metres will be drilled around the periphery of the Creek's moly deposit. This drill program has the potential to add considerably to the resource and provide crucial information for future development planning.

Presently, it is known that the deposit extends beyond the south wall of the proposed pit, as defined by Golder Associates Ltd. The holes are being drilled for exploration purposes and will provide insight into the size, shape and grade of the deposit well beyond the current proposed open pit.

Now that this work is progressing, Adanac is looking forward to Ruby Creek's future. "We expect to have our permits by late 2007—it's a tedious process." The proposed mine is also in the late stages of its B.C. environmental assessment. "And in that time, we should have the bankable done, all the detailed engineering, which is about a $15 million bill," Larry Reaugh explains. The procurement at this stage of "long lead time" major pieces of plant and infrastructure is critical to maintaining the aggressive schedule needed to complete construction by the end of 2008.

Presently forecasting a mine lifespan of 22 years, the Adanac Chairman notes that he can see that span expanding. "Probably, it will be up to 40 or 50 years. There are a lot of resources that we're actually discovering in further drilling and, if the price of moly stays strong...." The average operating cost for the first five years, is expected to be $5.87 US/lb Mo.

Aside from Ruby Creek, Adanac has several other moly prospects, of lower priority in comparison but still on Larry Reaugh's radar. "They're at the stage of historical resources, with the exception of one, which we've had a 43-101 done on, and we have an inferred and indicated resource on that one—the B&C Springs. That's a copper-moly. So, it's a little different." The three deposits, B&C Springs, Pine Nut and Cucamunga are located south of the border, in the state of Nevada. Not afraid of going abroad in search of a great property, Reaugh admits, "No, we could go anywhere. For example, I'm heavily involved in China in gold exploration; that's with another company." The B&C Springs copper-molybdenum deposit is approximately 200 kilometres southeast of Reno, in an expansive area that was extensively explored in the 1960s, 1970s, and 1980s by US Smelting Refining and Mining (USS RAM) sister companies, among others. The Pine Nut molybdenum deposit is in close proximity, about 80 kilometres south of Reno, near the Nevada-California state border. Nearby, a tungsten mine operated in the 1940s and Climax explored for moly in the 1970s. The Cucamunga molybdenum deposit is roughly 180 kilometres south by southeast of Reno, nearer to California. Adanac Molybdenum Corporation is steadily advancing the trio of projects with compilation reports, historical data confirmation drilling, and resource/reserve definition drilling to progress the evaluation, development and operation of mines at these sites.

Currently, the B&C Springs property is thought to be blessed with a bounty of 314 million pounds of moly. Prior drilling (66 diamond and RC holes) by USS RAM encountered molybdenum, copper and silver values. The discovery hole located in proximity to Adanac's claims averaged 0.30% Mo (0.50% MoS2) over 24.4 metres. Part of the tabular structure containing the discovery hole is located on a claim which has been retained by a subsidiary of USS RAM. Adanac has initiated discussions with the owners of the one claim in order to advance the project, which, based on the published grade and tonnage from the 1985 US Bureau of Mine circulation Principal Deposits of Strategic and Critical Minerals of Nevada, is reported to contain 131 million tons grading 0.12% Mo (0.20% MoS2). To date, however, Adanac has not been able to obtain enough of the original data to verify the classification of the resource or reserve. So, this historical estimate should not be relied upon.

The Pine Nut, near Carson City, Nevada, is estimated to hold 98.4 million pounds of molybdenum. According to the US Bureau of Mine data, it is stated that there are 82 million tonnes grading 0.6% Mo. The Pine Nut was diamond and rotary drilled by American Metals Climax Inc. in the mid-1960s. The Cucamunga deposit is described as 65 million pounds of moly. Past exploration consisted of drilling 52 diamond drill and rotary holes totaling 31,000 feet in the 1960s and 1970s by Bear Creek Mining, Molycorp, Geochemical Surveys Inc. and Duval Corp. The generations of exploration drilling have partially outlined two molybdenum deposits on the claims, the Basalt Cap and Roper Tunnel Zones. The Basalt Cap zone contains a drill-indicated resource of approximately 30 million tonnes grading 0.11% MoS2 (0.066% Mo).

In the words of Larry Reaugh, however, "Nevada will have to wait while we're putting Ruby Creek into production." It's crystal clear where his priorities lie. He adds, strategically, "The Nevada properties could be joint ventured, or we can just wait until we complete our work up at the Ruby Creek, and then we'll have some excess cash to start developing those properties."

Reaugh nods, "That's our game plan. Of course, our priority is Ruby Creek. It's not an easy job to get a mine into production. So, you can't squander your manpower and expertise on other areas while you're trying to do that, because it's an ongoing process that you've got to devote a lot of time to. And so, that's our major thrust."

Naturally, he says, "The other thrust is to develop the properties in Nevada and probably start production there, as well. And certainly we don't close the door on looking at new properties. There could be an opportunity that is just too great to pass up, or if, for example, a company that had an operating moly mine wanted to sell it; we'd be very interested in that."

With this assertive approach toward growing and expanding, which focuses principally on the moly mining angle, it's not surprising that Adanac is positioned as something of a leader in moly. "Actually, I see us positioning ourselves as a producer of molybdenum in North America. I would think we'll be a major producer," says Reaugh, "however, I wouldn't try to outdo Phelps Dodge with their operations in the US, for example, or maybe even Thompson Creek at this time. But, I think, ultimately, our main goal is to become the first new mine in the last two decades—at this size, open pit, 20,000 tonnes a day."

Adanac Molybdenum's past and future hinges on those Larry Reaugh describes as "key people" —the leaders within the company who are showing the drive required to actively ensure Ruby Creek's expectations of success. From aforementioned President and CEO Michael MacLeod to the company's "diverse group of directors," Reaugh says all have had an impact on getting Adanac where it is today: a renowned force in moly exploration and development.

"Robert Pinsent is the project manager on geology. And he was actually involved in managing the project for Placer back in the late 1970s. So, he's seen the continuity of it." Pinsent, Ph.D., P.Geo., was also the Qualified Person for the purpose of Adanac's National Instrument 43-101. "Then there's Rick Alexander and Mike Petrina."

Rick Alexander, P.Eng., was added to the expanding Adanac management team as Vice President of Project Development. Mr. Alexander is a registered professional engineer in B.C. with over 25 years experience in the development and operation of mining projects. Since graduating with a B. Sc. in Mechanical Engineering from the University of Alberta Alexander has directed all phases of project development, from feasibility studies through to managing detailed engineering procurement and construction management activities. His background includes international experience in Russia, the former Soviet Union, Central and South America, and Australia; and extensive experience in Northern British Columbia, Alberta and the Northwest Territories. He was instrumental in the development of Adanac's feasibility report for its Ruby Creek Molybdenum Project.

Mike Petrina joined Adanac as General Manager of the proposed Ruby Creek mine. He is a professional mining engineer registered in British Columbia, and holds an MBA from the University of Athabasca. Having spent over 25 years in the mining industry in open pit and underground operations doing everything from engineering and supervision, to project development, to consulting and contracting, his experience is considered a great asset for the development and operation of an open pit mine and concentrator.

Several more professionals are expected to be hired in the coming months, to assure Adanac has the ongoing leadership, experience and capabilities to build and operate the Ruby Creek mine. These additions are consistent with Adanac's strategy of adding expertise in engineering design, feasibility, financing, construction, and mining operations to its board of directors over the past twelve months.

Directors like Roger Taylor, P.Eng, F.I.M.M., a mining engineer with 45 years experience, have held senior positions in many operations. Taylor also has plenty of experience with acquisitions and feasibility studies, which were done for various major mining companies. From 1975 to 1979, Taylor served as Mine Manager of the Granisle mine (open pit copper-gold) and was also President of Zapata Granby Mining Corporation. He served as a Vice President of Cominco Ltd. and was President of their copper division from 1980 to 1985 where he handled the acquisition of Bethlehem Copper Mines Ltd. and Valley Copper Mines Ltd., to combine the assets of these companies under Cominco Ltd.

In addition, Taylor negotiated the permits and smelter contracts for the initial development of the Highland Valley ore body and handled the expanded Bethlehem mill to 25,000 tonnes per day to treat the Highland Valley ore. In 1984, he opened negotiations with Rio Algom for the eventual pooling of all the Highland Valley assets. Taylor has also held directorships in various other TSX-listed companies, as well as being a Mining Consultant employed on due diligence and evaluation of major mining operations in North America and Africa on behalf of mining companies and international banks.

Dr. David Stone is another mining engineer notably on Adanac's board. His career spans 25 years of consulting to the metal mining industry. His principal expertise is in mining rock mechanics where Dr. Stone has provided designs and operational advice for both open pit and underground operations worldwide. He has also authored a number of pre-feasibility and feasibility studies based on his broad knowledge of current underground mining methods, capital and operating costs, and project development requirements. Dr. Stone is President of MineFill Services, Inc., a mining consultancy based in Seattle, Washington with offices in Vancouver and Toronto. He also sits on a number of other boards of TSX-listed companies.

Neil Seldon is a Director with extensive experience in minerals and metal marketing. Following graduation from the Royal School of Mines in London, England, he founded Neil Seldon & Associates Ltd. (NSA) in Vancouver, B.C. in 1990. NSA act as consultants and marketing advisors for mining and metal industry clients in Canada, the USA, Australia, Chile and other countries around the world. Over the last sixteen years, NSA have provided advice and services to about fifty companies and organizations.

Ed Lee is a Director and Executive Vice President of the company. He has been an entrepreneur in private business in Northeast British Columbia for 20 years, twelve of which have been in the assistance of corporate development and finance for public companies. He is also currently the President and CEO of Reaugh's MolyCor company.

Chief Financial Officer, David Kwok has seven years of experience in the financial markets and oversees the financial and corporate functions at Adanac. His duties involve the administration of corporate and fiscal affairs, plus dealings with investors and institutions. Kwok has a Diploma of Technology in Accounting as well as a Bachelors of Business Administration from the British Columbia Institute of Technology.

Then there is Teresa Piorun, the Corporate Secretary, who has been with Reaugh's group of companies for 20 years. She is a senior corporate officer with wide-ranging responsibilities who serves as a focal point for communication with the board of directors, senior management and the company's shareholders. Piorun is also the advisor to the CEO, and other members of senior management, especially on corporate governance affairs.

"Overall, Adanac's head office has a staff of about fourteen, and during the season, of course, we have about 100 engineers working on detailed drawings through AMEC. We probably have another dozen engineers from other firms doing work for us. These are all contracts. And during the drill season, we'll probably have another twelve employees," notes Reaugh.

With the company's development of its Nevada moly properties temporarily on standby, it's 'all hands on deck' at Ruby Creek; maximizing the manpower at this pivotal time. A smart move, if you share Reaugh's point of view for future moly prices. "Our outlook is that you're going to see ever-increasing prices for molybdenum. Well, to be very honest, you may get $60, $70 moly in the next few years. It certainly should be $40 in the very near term here." Which makes for an exciting climate for constructing a pure moly mine.

"I see pandemonium in the total metals market," Reaugh continues, "specifically in molybdenum. And it's going to take years and years for the mining companies to complete their permitting, engineering and building of these projects to actually bring the price down. It's just that, there are no big projects out there that are far enough advanced to start immediately affecting the price. It's way out there in the future.

"That's driving the price up. And in order to catch up, there's got to be a few mines put into production. But then, obviously, you have the continuing growth in demand," Reaugh says in reference to the industrial revolution in Asia that he expects will strain existing moly resources for the next several decades. "Currently, the abundance of existing molybdenum deposits will easily fill increased demand, but they are still years away from obtaining production. If India kicks in, the same as China has, and they expect it will—they think it's going to happen in 2010, but I think it's already happening—you know, by 2010, I think they'll be at the same growth rate as China."

Logically, the seemingly never-ending demand for steel is often on the minds of those in the moly business, and Reaugh is no exception. "Because it's the steel industry putting real strains on the supplies," he says. "It just works its way through everything that's used in steel. And, well, you see copper is trading at historical highs, zinc is trading at super historical highs. Lead is close to a dollar, cobalt's at over $30 a pound. Manganese has jumped 300% in the last six months, and that's a crucial element of the steel and iron industry. You can't do anything without it. And so, I believe that it's just going to work its way through everything, ripple through everything."

With Ruby Creek's impending construction, Adanac appears poised to surf this ripple effect to its potentially fruitful conclusion. Moreover, time is on their side to achieve Larry Reaugh's bold dream to "become the first, new, large-scale, open-pit, primary molybdenum producer in 25 years."

BARD VENTURES LTD.

President and CEO of Bard Ventures Ltd., Eugene Beukman, is banking on molybdenum to take this rising junior exploration company to the next level. With impressive business acumen and a sharp legal brain, the South Africa-born former lawyer brings over twenty years experience in the acquisition of assets and joint ventures to the table. It is an ideal background for this industry, and one that lends credibility to Beukman's game plan to aggressively acquire moly properties to add further momentum to Bard Ventures' growth.

Naturally, Beukman's words are those of a lawyer—smart, succinct, and sophisticated. He speaks with an exotic South African twang as he explains how he became involved in Bard Ventures, and how the company acquired its key property, Lone Pine, which is pure moly.

"Well, I came to Canada about 14 years ago. I used to practice law," states Beukman, who graduated from the Rand University of Johannesburg with a Bachelor of Law degree and a Bachelor of Law Honours Postgraduate degree in 1987, and who used to be an Advocate of the Supreme Court of South Africa. "I worked as a legal advisor for the South African mining house Gencor. We handled the 'unbundling' of its operations." Gencor would merge with Billiton (a British mining company with a South African background), and Billiton with BHP (an Australian company) to ultimately create the largest mining firm in the world.

"At this stage, I decided to immigrate to Canada and move permanently to Vancouver. I started consulting. I met up with a group of people and we eventually began managing public companies here. I started Pender Group Consulting with a partner. Essentially, we take care of all the regulatory filings for our clients, we look after all their stock exchange requirements." Beukman continues, "We also help them do SEDAR filings, arrange the Annual General Meeting (AGM) material, and draft financial statements reviewed by the auditor. At the end of the day, we prepare the AGM material, liaise for the transfer agent, we call the AGM and hold the meeting here in our offices. Now, one of these companies that we manage is Bard Ventures. I focus more than 50% of my time on Bard Ventures ,looking specifically at the moly properties that we have."

Bard Ventures, listed on the TSX Venture Exchange (symbol: CBC), was formed 20 years ago and Eugene Beukman has been CEO for over 10 years.

Focused on mineral exploration and development, the company has amassed a generous and varied portfolio of natural-resource properties ranging from advanced exploration properties, to those at a grassroots prospecting and mapping stage. All the while, as Beukman stipulates, moly has been front and centre.

"Lone Pine is moly," he says of Bard's premier property located in the Omineca Mining Division, just outside of Houston, British Columbia. Bard Ventures entered into an option agreement in August 2006 to acquire a 100% interest in Lone Pine. The company is presently exploring for porphyry hosted molybdenum mineralization. The property area extends over several molybdenum showings—Quartz Breccia, Alaskite Zone, Mineral Hill and Granby—that have been previously documented in various assessment and government reports.

"We tested some of the showings last year, but we're now at this stage where we're doing the geophysical surveys, so we can get a signature of what's going on. We have planned and secured a drill rig from the end of June through December 2007, because drill rigs are so in demand these days, you know." This summer drilling program is a follow-up to Bard Ventures' seven-hole winter 2007 diamond drilling program, which confirmed the presence of widespread molybdenum mineralization on the property. "Our present results are running at 0.09% MoS2 which we think is mineable," concludes Beukman. The winter program predominantly targeted the Quartz Breccia and Alaskite Zones and returned significant molybdenum intersections in all drill holes. The best hole returned an average of 0.07% MoS2 over the entire 489.4 metres, including a higher-grade intercept of 231.4 metres of 0.093% MoS2. Another drill hole, located midway between the two zones, intersected a high-grade interval in excess of 12 metres at 0.27% MoS2.

SJ Geophysics Ltd., of Delta, B.C., mobilized a geophysical crew to the 3,155 hectare property in May 2007. An approximately 55 kilometre 3D IP (induced polarization) survey is underway, in conjunction with a ground magnetometer geophysical survey. "We've got to complete the geophysics," Beukman explains. "As it's being done we can interpret the data on a daily basis because all the information is downloaded every night and sent to our geophysicists in Vancouver. So we don't have to wait till the survey is done before we get the data.

Once the geophysical program is completed, all historic data and the winter 2007 drilling information will be digitally compiled for targeting the higher grade molybdenum mineralization. The services of Foraco

Drilling Ltd. (formerly Connors Drilling) have been contracted. "We're going to see if we can drill, to start with, 20 to 25 holes, and I've planned for at least 45 holes on the property—this is the moly property we've got to focus on," confirms Bard's Beukman. "This is quite an extensive program, so we could be spending close to $2 million on this property in the next six months."

Summer is peak-season at the Lone Pine site. "We have up to 10 people working in the field, as well as a geophysical crew of about nine guys. We have our line cutters going in front of them—the guys doing the GPS survey. Once that is followed up by some prospecting and interpretation, then your project geologist runs the project. He makes sure the drill contractor gets all the work done on time and in the right locations," says Beukman. The Lone Pine exploration work is being conducted under the supervision of a Bard Ventures Director, Professional Geologist James Miller-Tait.

One of the high priority areas on the property that will be tested is the Granby Zone, located approximately 700 metres to the east of the Quartz Breccia Zone. In 1976, Granby Mining Corporation drilled 682.8 metres in 12 short percussion holes into the Granby Zone. Vertical percussion hole number 'M3' intersected 0.176% MoS2 over 36.6 metres from a downhole depth of 24.4 metres to 61.0 metres (final hole depth) and this intersect included a higher grade interval of 0.30% MoS2 over 18.3 metres, between 24.4 and 42.7 metres.

Historically, mining and exploration have been conducted on and around the area of the Lone Pine claims since early in the last century, with a considerable amount of geological, geophysical, and geochemical work having been done on the property since 1976. Over the decades, many different companies have taken part in exploration on the property, including Canex Aerial Exploration Ltd., Molymines Exploration Limited, Cominco Ltd., Granby Mining, and Noranda Mining and Exploration Inc.

Several programs of diamond and reverse-circulation drilling have been conducted since the mid-1970s, the most notable of which was the program carried out by Granby Mining Corp. in 1978. During this program a hole drilled in the Quartz Breccia Zone returned 356.3 metres of 0.068% MoS2 from 20.7 to 377 metres which included a higher grade intercept of 154 metres of 0.088% MoS2 from 181 to 335 metres. Additionally, a second, deeper diamond-drill hole in the Alaskite Zone returned 343.7 metres of 0.06% MoS2 from 3.6 to 352.3 metres, including 101.4 metres of 0.078% MoS2 from 3.6 to 105 metres. Both of these drill holes were terminated in mineralization. This mineralization was confirmed by the drilling undertaken by Bard Venture's in the winter 2007 program.

Adding to the viability of this property, logistically Lone Pine could not be better positioned. "On the property, we actually have a B.C. Hydro sub station, which is a high voltage power transmission line. Highway 16 runs through the property—it's fairly flat. And there's a natural gas pipe line on the property as well. Now if you try and compare this with other companies doing this kind of grassroots exploration work, they're in the middle of the bush, they've got to fly somebody in with a helicopter, you have a diesel generator that runs your camp and your drilling. Whereas we basically 'plug-in,'" Beukman laughs.

"Also, we're only 11 kilometres from the CN rail line in Houston, and about 40 kilometres from Thompson Creek. We have a 4,000 hectare property here and we recently acquired an additional 3,000 hectares with our Grouse Mountain property. So, we really have a very large area," he says.

The existing infrastructure makes the sizeable Lone Pine and its prospective wealth of molybdenum even more attractive to Beukman, who believes the project can easily and rapidly be moved forward to development. "The drilling is moving us towards this goal." In addition, Bard's newly acquired Grouse Mountain property, which neighbours Lone Pine, is making advancements of its own. "We're doing some line cutting at the moment, and we'll do a geophysical program," says the Bard President.

The seven claims that comprise Grouse Mountain—immediately to the north and adjacent to Lone Pine—contain 94 cells in total and are wholly situated in the Omineca Mining Division. The property covers a number of polymetallic showings, the most notable of which are the Copper Crown, Rainstorm, Eureka, North Lake, Hidden Treasure, Solo and Shorn. The geology of these claims is favourable for porphyritic molybdenum mineralization.

Exploration on the Grouse Mountain property actually dates back to 1914 when chalcopyrite/sphalerite mesothermal vein mineralization was first discovered. During the period of 1915 to 1927, extensive drifting and raising was completed on two levels on the property's Ruby Zone, with some work being done on its Lakeview Zone as well. Then, after a roughly 30-year hiatus, a mining company completed 5,700 metres of diamond drilling and rehabilitated the underground workings.

In1970, VLF-EM (very-low-frequency electromagnetic) surveys and 1,282 metres of diamond drilling was completed. This was followed by various geochemical/geophysical surveys and supplementary drilling in 1984, when the operators added a further 720,000 tonne low grade inferred resource to the existing 320,000 tonnes of 0.38% copper, 4.23% zinc, and 0.88 g/t silver. (Note: These resource calculations were completed prior to a National Instrument 43-101 report, and therefore are non-compliant with current regulations.) A short exploratory drill program in 1990 is the last known major work done on the Grouse Mountain property.

Adding to the potential of this property, there are a number of mineralized structures on the Grouse Mountain property and to date only a few of them have been tested by drilling. Several geologists have hypothesized that porphyry style mineralization may also exist at depth here. Bard Ventures plans preliminary geophysical, geological, and rock sampling work to verify the historical data. This will be followed by an exploratory diamond-drilling program of its own.

Beukman adds neither Grouse Mountain nor Lone Pine has any expected issues with Aboriginal claims. "We haven't received anything at all at this stage. Our process has been to liaise with the regional authorities, and look to the mine's inspector and for guidance. What happens is we usually enter into an agreement where we provide employment for the local people and, you know, try to work together." Beukman once again demonstrates his diplomacy; surely a preferred trait in every good lawyer.

Along with Bard Ventures' main moly prospects, the company boasts a diverse portfolio of other natural resource properties. Also in B.C., the Ingenika/Swannell and the Wasi Creek properties boast zinc, lead and silver.

Ingenika/Swannell—optioned in September 2004, with a 50% interest for Bard—is also in the Omineca Mining Division, located in the Swannell Range of the Omineca Mountains 103 kilometres north-northwest of Germansen Landing. Because of intense logging activity in the area the property shares with Lone Pine the same comforts of excellent access—in this case via a main logging haulage road.

There have been numerous exploration programs completed on the property, focused on the site's Ingenika, Onward and Swannell Showings. A resource estimate, completed in 1969 on the Ingenika Mine, included 22,677 tonnes grading 119.9g/t silver, 9.8% lead and 6.1% zinc. Grab sampling at the Onward Showing trenches assayed up to 64.2% lead and 1,870g/t silver, and at Onward South, grab samples from an old shaft assayed up to 25.62% zinc, 16.28% lead and 139.8g/t silver. Drilling by Selkirk Metals Corp., with whom this is a joint venture, in the Swannell

Showing revealed 5.82% zinc, 3.17% lead and 28.2g/t silver over 3.0 metres; 14.5% zinc, 4.3% lead and 37.7g/t silver over 1.7 metres; and 5.83% zinc, 1.29% lead and 15.3g/t silver over 9.6 metres.

Selkirk is also jointly involved in the Wasi Creek property, optioned by Bard in September 2004 (50% interest), with Bard currently funding exploration and Selkirk operating of the program. Wasi Creek is located on the south side of the Osilinka River, 150 kilometres northwest of Mackenzie, B.C. It covers areas of historic trenching and diamond drilling for zinc-lead-silver mineralization in the Cambrian Atan Group in the north of the province.

Wasi Creek was first discovered and evaluated by Cominco, in the early 1990s. Cominco identified zinc-lead-silver mineralization along a two kilometre carbonate horizon known as the Par Horizon. In 2002, Cross Lake Minerals Ltd. discovered the Carrie Zone at Wasi Creek, a new zone of mineralization parallel to the Par Horizon. Initial hand trenching by Cross Lake discovered sphalerite-galena-pyrite mineralization that assayed 5.10% zinc, 0.89% lead and 18.0 g/t silver across a sample width of 10 metres. Work done by Bard and Selkirk since 2004 has focused on these two mineralized zone.

A 1000-metre diamond drill program carried out by Selkirk Metals in 2005 targeted a new horizon, sub-parallel to the par horizon, which had been initially identified by airborne magnetic and EM geophysical surveys. In addition, the geophysical survey results identified several high priority targets that warrant further ground examination and follow-up exploration including mapping, prospecting, rock and soil sampling, trenching and, depending upon these results, more diamond drilling. One high priority area is the southern extension of the Par Horizon which includes several targets. For instance, in 1992, Cominco intersected 6.35% zinc, 1.07% lead and 16.5g/t silver over 4.8 metres and 4.98% zinc, 0.68% lead and 18.0 g/t silver over 5.0 metres in a hole north of the high priority area.

Further trenching of the Carrie and the Par Horizon by Selkirk Metals in 2006 successfully extended the area of known mineralization on the property and delivered favourable results. The main focus of this 2006 program was the Carrie showing, and the trenching program uncovered 5.68% Zinc over six metres. Diamond drilling, undertaken late in 2006, intersected 5.61% zinc, 0.41% lead and 7.2 g/t silver over a width of 1.45 metres from a downhole depth of 196.25 to 197.70 metres. The mineralization in this intersection, located at the northern end of the Par Horizon consists of honey-coloured sphalerite, galena and pyrite. Ultimately, the data from the 2006 trenching and drilling programs will serve to guide further exploration of the Wasi Creek property, including an additional program of diamond drilling.

Also of interest at Wasi are numerous large, angular, higher grade, massive sulphide boulders. The source of these boulders is unknown. Two of these boulders, sampled by Selkirk, returned significant assays. One contained 26.30% zinc, 25.98% lead, and 96.3 g/t silver in one, and the other 8.46% zinc, 42.43% lead and 384.8 g/t silver.

Therefore, Bard and Selkirk are planning to conduct further exploration on the Wasi Creek property—including trenching, prospecting and geological mapping. Once target definition has been completed, diamond drilling would be undertaken to test priority targets. Jim Miller-Tait, P. Geo, is supervising the work; as well as being a Director of Bard, he is also Vice-President of Exploration for Selkirk Metals Corp.

So, overall, the prospects look promising for Wasi Creek and Ingenika/ Swannell; just two of Bard Ventures' emergent projects. Bard's portfolio also includes properties outside of British Columbia, such as the Opikeigan Lake property, in the Thunder Bay Mining Division, 10 kilometres northwest of Fort Hope, Ontario. A 50% interest in Opikeigan was acquired in February 2006, in another joint venture, this time partnering with SLAM Exploration Ltd. The Opikeigan property hosts four gold occurrences along with the historic Fort Hope Gold Mine.

Previous drilling at Opikeigan, conducted in the late 1980s, returned several intersections with gold grades as high as 113.87 g/t Au over 0.1 metres, 28.83 g/t Au over 1.5 metres, 6.2 g/t Au over 7 metres and 19.9 g/t over 1 metre. Bard and SLAM are planning to drill a gold-bearing zone that lies at the centre of a gold mineralized system on the property.

"Currently, we have a zinc property, a gold property—joint ventures, both of them—and those are things we'll spend $200,000 on and have small drill programs. But the focus is definitely on the moly property. "With moly being in excess of $30 per pound—and when Lone Pine was previously drill tested it was a mere $2 per pound—that's over 1000% better," emphasizes Beukman. "We've got to get in there, take advantage of this and see, with the funds available, if we can have some resource identified there, which we feel comfortable we will be able to accomplish."

With a concerted drive to reach the next plateau Bard stands out from many of its rivals. "We're trying to get to that next level, and we are fortunate in this company that there is a lot of focus on it," Beukman says, ambitiously. "In the meantime, we will try to aggressively grow by acquisition as well. We have been talking to a few other smaller companies and seeing if there is someway that we can enter into joint ventures or take over some of their properties, with the focus totally on moly."

Certainly, the possible acquisition of smaller moly companies depends mostly on finding a company that is not interested in becoming part of the moly-mania phenomenon, which is no easy task. "We have entered into discussions with two companies at this stage, and they have taken it to their board—but, usually, they're all sitting around waiting for the biggest price they can get, because molybdenum is in such high demand at this stage."

Funding is one thing that Beukman's company isn't short of. "We're fully financed. We've got $6 million in the bank. We have some outstanding warrants, which would give us another $2 million. And, we're a big volume trader—we can trade 10–12 million shares a day. The last few weeks were slow for us, but we still averaged trading 400,000 to 500,000 shares a day—whereas other junior companies really struggled to get 10,000 or 20,000 shares going, you know. So, we're fortunate in that case. But we now have to do the exploration work, get the results out, and get to the next level where we can trade in the $1 range. Then, immediately, you get on to the radar of other people, and they follow your company more closely." Beukman continues, "We believe that in the foreseeable future, we'll be in a different league."

When heading into the big leagues, you need a major league team of big hitters. At Bard Ventures, John Malysa, Jim Miller-Tait and Emmet McGrath work alongside Beukman, providing years of professional skills and experience.

John B. Malysa is a mining engineer with an extensive background in the mining and exploration environment. As a Registered Professional Engineer in Colorado with a B.Sc. in Mining Engineering from Penn State University and a MBA from the University of Colorado, he has amassed in excess of 30 years progressive mining experience in all aspects of both surface and underground mining—exploration, design, feasibility, construction, operations and management.

Malysa, who was appointed to the board of directors of Bard Ventures in March 2007, also has 'hands on' underground and surface mining experience in North and South America and in both union and non-union operations. His project and operations experience include underground mines up to 22,000 tonnes per day and surface mines up 100,000 tonnes per day. Additionally, he has management, design and construction experience in various precious- and base-metal projects with new mines costing upwards of $250 million (USD). His mineral process knowledge includes crushing and screening, CIL, CIP, heap leach, gravity and flotation recovery methods. Malysa has participated in financial and union negotiating with banks, investors and employees. And he brings to the table a proven track record in holding several positions as President and/or General Manager of entrepreneurial mining companies, reporting to their Boards of Directors.

Jim Miller-Tait, a Director at Bard responsible for overseeing exploration at Wasi Creek and Lone Pine, has held numerous positions throughout his career, including Vice President of Exploration for Selkirk Metals Corp., Vice-President of Exploration for Cross Lake Minerals since November 1998; Project Geologist, Cross Lake Minerals Ltd., from January 1998 to November 1998; President, Sikanni Mine Development Ltd., January 1997 to January 1998; Consulting Geologist, May 1996 to December 1996; Project Manager (previously Chief Geologist), Oniva International Services Ltd., September 1987 to May 1996. "He's a very experienced geologist," Beukman points out.

Another Director of Bard, Emmet McGrath, is an independent chartered accountant. He also serves as a director for Cross Lake Minerals Ltd., a company listed on the Toronto Stock Exchange, and Selkirk Metals Corp., a company listed on the TSX Venture Exchange. He graduated from the University of Calgary in 1971 with a Bachelor of Commerce Degree. From 1981 to 2002, he was employed as a partner at KPMG for Greater Vancouver.

With such an accomplished and capable team, Bard Ventures is well prepared for their continuing ascension into a league of their own. And with moly properties at the forefront of his agenda, Eugene Beukman is determined to ensure this happens.

COLUMBIA YUKON EXPLORATIONS INC.

Ronald A. Coombes and Douglas L. Mason are the men behind mineral exploration outfit Columbia Yukon. With Coombes as President and Mason as Chairman, the company is engaged in the acquisition and exploration of precious and base metal properties located specifically in British Columbia.

"B.C. is our main area," says Coombes, assertively. "Of course, Columbia Yukon's "Storie Property" is the moly deposit that we're moving forward on, to try to see whether or not we can get a world-class situation out of it." The Storie Property is a molybdenum deposit in northwestern B.C., offering the potential of 100.5 million tonnes at 0.077% Mo (0.129% MoS2), based on a historical resource estimate (which historical resource estimate has recently been validated by a NI 43-101 Mineral Resource estimate report). The company is making swift progress at the site, which it acquired in May, 2006. "We've actually started drilling there. We're going to be drilling somewhere between 30,000 and 40,000 metres this year. It's a big drill program," Coombes says.

Ron Coombes is appropriately qualified for his presidential duties at Columbia Yukon. A Vancouver businessman who founded Orphan Boy Resources Inc. in 1995, Coombes is an experienced entrepreneur and fundraiser who has been involved in mineral exploration and junior exploration company management for the past decade.

"Orphan Boy is an historical name related to a famous Crown grant that was located in the McCulloch Creek area of B.C. It was pretty important at one moment in time; there were 10,000 people working in that particular area of British Columbia looking for gold deposits. The name has since been changed to International Bethlehem Mining Corporation (IBC), but, yes, I am still the President and CEO of that company," notes Coombes, who says that any suggestion that he is "pretty busy" is an understatement. He is also President of Venturex Explorations Inc., making a hat-trick of mining companies for Ron Coombes and Chairman Doug Mason.

"We're focused on all three companies. Logistically, I spend a lot of my time on the other two companies as well as on the Columbia Yukon Storie deposit," explains Coombes. "However, what we do have at this time is more consultants working on the Columbia Yukon Storie deposit. We've assembled a great team of senior geologists to work on all of the companies' projects."

Doug Mason adds, via speaker-phone from Columbia Yukon's offices in Vancouver, "We were fortunate enough to have this Storie moly deposit come our way. We entered into an option agreement to place it into Columbia Yukon and I was introduced to Ron Coombes here, who is an excellent president. It seemed like a very good reason to come out of my 'semi-retirement' to focus on what I believe could be one of the largest homeruns of my twenty-plus year career in the public marketplace."

Former President and CEO of the Clearly Canadian Beverage Corporation, Mr. Mason has enjoyed many successes in the public and private financial arenas over the years. Currently, aside from his chairmanships at this trio of exploration companies, Mason holds the titles of President and CEO of Waterfront Capital Corporation and President, Director and sole shareholder of Criterion Capital Corporation, a private investment and financial consulting company.

"I got into the public market game back in about 1985, with a beverage company called the Jolt Cola Company. That was my first public company play and that turned into what became, in 1987, the Clearly Canadian Beverage Corporation," says Mason. "The Clearly Canadian Beverage Corporation was a stock that went to $500 million market cap. And, after twenty years at the Chair position, I decided to resign in the middle of 2006 to focus on mining. "Now, I got involved with mining back in the early 1990s. I started a company called Commonwealth Gold Corporation and we were lucky enough to be the third company that discovered diamonds in the Lac de Gras," a lake approximately 300 kilometres north of Yellowknife, Northwest Territories, which was the centre of the Canadian diamond rush of the 1990s.

"Under Commonwealth Gold Corporation, there was a merger offered by a company called Aber Resources. Aber and Commonwealth merged to consolidate the Lac de Gras claims—half of them. The other half of the claims I had in that area I joint ventured to Winspear Resources, which ended up being bought by De Beers. But, in any case," Mason continues, "at that point I became the largest shareholder of Aber on the consolidation. So, that was enough to get me very interested in the mining segment. I bought another company, and that's when I started the Columbia Yukon organization.

"Our first major project was chasing after a large nickel deposit. As we rolled into the mid 1990s, I believe we were one of the largest landholders in the whole of Labrador—we had about 18,000 acres of claims there, and we spent about $15 million looking for another nickel deposit."

"We still do have, in our group of exploration projects, an exploration property in Labrador that has three drill intersections of significance—nickel, cobalt and copper," notes Ron Coombes of a Voisey's Bay-based project owned by Columbia Yukon along with International CanAlaska Resources Ltd. A 1997 drill program on the property, comprised of seven holes for a total of 9700 metres, delivered encouraging results. "We may get back onto that project later. Or we could possibly joint venture it."

"Yes, one initially good aspect was that the Labrador claims were contiguous to the Cartaway Resources play," says Mason. "We announced some very good nickel intersections and then, about a week later, the Cartaway Resources' scandal kind of blew up and it diffused the whole area play and the idea of looking for nickel at that time in Labrador, needless to say."

"Unfortunately, Cartaway Resources made some statements and put out some results that were not true," explains Coombes. "Of course, the public takes exception to that, especially since the Bre-X fiasco that happened not too long before the Cartaway situation. Anyway, our Labrador property potential is real. So, we're looking at resurrecting that particular property and moving forward to finding either a joint venture partner, or we'll do some work on it ourselves."

Regardless, "I've always been involved in a few interesting mining companies," says Doug Mason. "We've always been exploring, mostly in Canada, with a little in Africa. And so, my career kind of evolved along with a variety of industrial and mining companies. I've raised about $250 million in my career for public ventures. Then, commensurate with my retirement from the beverage industry, and kind of quasi retirement in my life, I once again got very interested in mining, just as the cycle was starting to pick up again."

When Mason met Coombes, things were—in Coombes' own words—getting "ugly" at Orphan Boy Resources. "I was in a bit of a debacle. In 2005, I had some serious questions that I wanted to ask with regards to a project that we had," explains Coombes. "I called one of the engineers to task on it and, in doing so, it got very difficult as he tried to eliminate me as the President of the company.

"I ended up having to call an Extraordinary General Meeting for Orphan Boy Resources in order to straighten matters out. And, at about that same time, I met Doug Mason. We sat, we talked. Doug was looking for somebody at that time to run his mining companies for him. We felt that the synergies were there."

Coombes smiles about the eventual outcome of their meeting. "Doug's a great guy. We both felt very good about our synergies. And so we joined forces and, you know, we let a few properties go. We had to make a number of changes that weren't all that popular. But, that's what happens when you reorganize. And then, of course, from there we've made some significant property acquisitions—the Storie moly deposit being one."

"Certainly, about two years ago, we got really interested in focusing on certain projects," reveals Doug Mason. "But, at this stage in my life, I was more interested in starting companies that had something already in the ground rather than grassroots exploring." Hence, Columbia Yukon's Storie deposit could not have arrived at a more opportune time for the semi-retiree and the company's President.

"Meeting somebody like Ron Coombes, who is market-savvy, I get to combine my twenty-plus years of experience in raising capital and making the public aware of certain plays with his," says Mason. "So far, we've raised about $20 million in the last eighteen months for Columbia Yukon. We're in very good shape, capital-wise. So, you know, we're extremely excited about the whole play. We believe we've got a real world-class property here that's got a very strong possibility of turning into a mine."

The possibly mine-worthy Storie moly deposit was acquired from a private Calgary–based exploration company, Eveready Resources Corporation, with the option for Columbia Yukon to earn a 100% interest in the project. The project was discovered on a golf course and introduced to the Company by Gary Kilpatrick, who was a former director.

"Shortly after we had acquired the property, molybdenum—of course—began rising," says Coombes. "All of a sudden everybody started getting on the bandwagon and talking about moly and how important it is. So, our timing couldn't have been better."

Located directly south of the former Cassiar, B.C. Mine town-site, the Storie Property is 125 kilometres north of Dease Lake and 145 kilometres south of Watson Lake, Yukon. It is benefited by highway access to the immediate project area and a two-wheel drive dirt road to the deposit, as well as being in close proximity to labour, telephone and an airstrip.

For over twenty-five years, the Storie deposit has been documented through exploration. It was most recently drilled by Shell Canada in the late 1970s and early 1980s. Shell estimated a resource of 100 million tonnes, grading 0.077% Mo (or 0.129% MoS2). Although this estimate did not meet current NI 43-101 standards, and Columbia Yukon's qualified person, John Kowalchuk, PGeo, had not reviewed the calculation for the historical resource, he believed that it has been calculated by competent engineers to the standards of the period.

Coombes states, "Right now, within the 100.5 million metric tonnes, there's a historical contained metal value of about 170 million pounds of moly. And we're hoping to increase that number through additional drilling and exploration."

Based on eighty-six diamond drill holes, covering an approximate 1000-metre strike length, over a 300- to 400-metre width and to a vertical depth of 150 metres, Shell's work indicated that the deposit is, geologically-speaking, open along strike to the east, north, west and to depth.

The history of exploration on the site dates back to 1959–1961, when Cominco drilled twelve holes. Casmo Mining drilled fifty-eight holes from 1964–1968. Levana drilled four holes in 1971. Between 1968 and 1975, Consolidated Coast Silver made fifty drill holes in the area and completed over 600 metres of underground development on a lead–zinc–silver discovery. In 1976, Balfour Mining completed trenching and diamond drilling. Next, from 1978 to 1981, was Shell Canada's extensive diamond drill program on the Storie moly deposit. Then Eveready Resources acquired the project in 1997. Eveready compiled data, trenched and rehabilitated roads up until 2003, before optioning it to Columbia Yukon in 2006.

These prior explorations revealed, through induced polarization (IP) and ground magnetics, that the mineralized system could be open for an additional 400 metres eastward. Also, random historical deep drill holes demonstrated that mineralization does continue in certain areas to at least 200 metres or more in depth. Thus, Columbia Yukon intends to redrill the Storie molybdenum deposit sequentially along strike and to an approximate depth of 300 metres on 100-metre drill centres, to fully delineate the deposit. The company figures that, given the size of the geophysical target and the location of mineralized drill intersections outside of the resource area, there appears to be potential to significantly increase the size of the deposit.

Ultimately, the exploration potential of the Storie moly deposit is based on the past data, which appears to show reasonable consistency of grade between drill holes. It is also noteworthy that float samples grading up to 0.50% MoS2 have been found to occur 500 metres south of the deposit, plus a 1968 hole reportedly drilled by Coast Silver on the "M Zone," located one kilometre to the east of Storie, graded 0.23% MoS2 over 130 metres.

Historical records show that there is little, if any, analysis of metals other than molybdenum, however, numerous silver–lead–zinc–gold replacement skarn-type occurrences are evident throughout the project area. Apparently fifteen are known of along the margins of the Cassiar Stock—these kinds of occurrences are common peripheral to large-sized porphyry molybdenum systems—perhaps indicating the Storie's significant moly potential.

With Storie's moly prospects; the Coombes/Mason team have plenty to be happy about. "Yes, we've got something right now that's already significant, but we believe it may very well be a lot larger than it already is. It's the sixth largest deposit in the Province of British Columbia right now, and we're hoping to bring it up to the number three or number two," Coombes says.

Columbia Yukon conducted a 2006 drill program of twenty holes (5000 metres) on Storie, designed to twin many of the holes that were drilled by Shell Canada, as well as holes drilled by New Jersey Zinc Co. – once the largest producer of zinc and lead products in the U.S. – in 1970 and 1971. The assay results were heartening with intervals reflecting a zone of mineralization greater than 0.04% Mo. All of the holes intersected molybdenite mineralization from top to bottom. The results are similar in thickness and grade to the holes that were drilled by Shell, appearing to confirm the historical resource calculations.

Now, Columbia Yukon's 2007 drilling program has been increased to between 30,000 to 40,000 metres. At an estimated cost of approximately $6 million, which, as Mason explained, Columbia Yukon has secured from recently completed financings. The Company presently has four drill rigs drilling on the Storie Property.

Previously 10,000 to 20,000 metres of drilling was planned for 2007, however the increased program is designed to further define and potentially expand the known historical resource and to provide additional confirmation of the resource with infill drilling. Also, the drilling is intended to test the potential to increase the size of the resource with exploration along strike and in parallel targets.

In July 2007, the company received a NI 43-101 compatible resource report from Watts, Griffis & McOuat Limited, Consulting Geologists and Engineers of Toronto, Ontario ("WGM"). WGM has also been engaged by Columbia Yukon to complete a scoping study for the Storie project, with a view to taking the property toward full feasibility.

Things are moving forward promptly and efficiently. Yet, there hasn't always been such a nurturing climate for 'wannabe' explorers and miners, notes Columbia Yukon's Ron Coombes. "We experienced fifteen years of probably the most horrific times in B.C. mining history. I mean, a previous government that was anti-mining and metal prices that were at historical lows. Those circumstances really made us wonder what we were doing. But, of course, things have turned around in B.C. now as it is a cyclical business.

"The cycle is normally seven to ten years. Well, we went through a fifteen year hiatus there, which was twice as long as what was to be expected for a turnaround." This naturally tested the confidence of even the most devout who were working in the mining game. "But, that's why we're very confident now," concludes Coombes. "Because of that hiatus the resources of existing mines are depleted. And to put a new mine into production, there is a significant time period involved. So, at this stage, we believe that we could very well see strong metal prices for some time."

Optimistic and enthusiastic about the future, Coombes continues, "It's looking very good." Nevertheless, he tempers his positive thoughts with a dose of realism. "Of course, one must always be conscious of the fact that the current trends may not continue forever. And you never know for sure when things may change because there are just so many variables that come into play with this business. You know, a new war all of a sudden gets announced. Or something crazy happens internationally. Currencies go up or down."

Aware of this, Coombes says, "But, yes, we're going very well right now. I think with our Storie moly project, our plan is to expand it, to make it as large as possible. Essentially, my philosophy is: 'expand, expand, expand.' Make it as big as it possibly can go."

"Exactly," adds Mason. "It's the right metal at the right time with the right deal structure, and the right property with the historical resource on it."

The Storie moly deposit practically speaks for itself, with all the makings of a great open-pit mine operation. "It really does; it made the decision very, very easy, when you're starting with that kind of an asset that's so potentially expandable and has so many favourable logistics," says Mason. "We certainly trust Shell's historical data. And, you know, we've had Watts, Griffis & McOuat working on updating the historical data into a NI 43-101 Mineral Resource estimate report so that it's now compliant with today's standards.

"But, in the meantime, we're planning on drilling between 30,000 to 40,000 metres this year, of course. We've also negotiated good arrangements with First Nations. We've got our environmental baselines on the way, and we've got a dozen geologists working on the team. Four of them are 'in-house' which is very helpful for us." Moreover, "having multiple mining companies in our camp helps us to carry the infrastructure, and not burden it on any one company—that's also a nice advantage for our group (International Bethlehem, Venturex and Columbia Yukon)," states Mason.

"We also have our own in-house legal, our own investor communications, our in-house mapping and staking, our environmental people, and our own prospectors. All of those things make it a lot easier to assist us to getting to a successful conclusion—which is making a mine. That's something I've never done, and something I've promised myself to do this time."

In addition to the Storie moly deposit, Columbia Yukon has its Barnes Creek gold–silver property, which lies at the headwaters of the Kettle River in B.C. It has a soil anomaly, best defined by gold and arsenic, that trends northwesterly and is approximately 1200 metres by 300 metres and open both to the northwest and southeast. Maximum soil values are 1280 ppb gold, 875 ppm arsenic and 8.9 ppm silver. Trenching at the eastern part of the anomaly – known as the Holmes Lake Target in the Spring of 2005 uncovered high-grade gold- and silver-bearing quartz veins with assays to 16.6 g/t gold and 282 g/t silver.

Columbia Yukon downplays this property somewhat as "Barnes Creek is really an early stage exploration project. Although, you always remain optimistic, this one is too early to get excited about. So, it's the Storie Property that is Columbia Yukon's primary focus," says Coombes.

"Yes, and we still have our potential nickel discovery back in Voisey's Bay, Labrador that we intend to get back to," confirms Doug Mason. "I've always pledged that, before I retire, we're going to get back there and drill that again. So, we might just have a nickel play as well."

Keeping their options open and diverse is undoubtedly the name of the game for the team at Columbia Yukon. "You bet," says Mason. "And we've got a number of other companies with some very good exploration results as well. We've got one company, the International Bethlehem Mining Corp., where we have a custom mill and some very interesting exploration projects. We're excited about that for the future. And Venturex Explorations Inc. looks like it could be very successful as well with its zinc–lead–silver property, which we refer to as the 'Carnes Creek.' So, we've got that, and we've recently made an option acquisition on 8000 hectares down in Mexico, also for Venturex.

"So, there's no sense laying down the swords with this much activity and appetite in mining in Canada," says Mason, who plans to stave off retirement for at least another few years.

Certainly, with the present levels of activity and the ongoing voracious interest in mining, one might feel added pressure from the competitive nature of the industry. Yet, Coombes says, "I never worry about other companies. They're doing their projects and we're doing ours. Sure, it's competitive when you try to locate and acquire properties, but the actual development and exploration—we're all equally enthused for one another. There's really not a competitive environment, when it comes to that point, once you've got your properties and you're developing them.

"I hope everybody is successful," he says. "That isn't going to happen, of course, but what is important is acquiring properties of merit—properties that are near existing roads, skilled labour and infrastructure. If you can reduce the risk in those three areas, you have a much better chance of success."

Success is also contingent on developing a skilled senior management team; something that Columbia Yukon has excelled in. With a well-rounded team boasting in-depth knowledge of financing and exploration management, along with a superior technical Advisory Board, Columbia Yukon is very well positioned to maximize its B.C. project portfolio potential, covering molybdenum, gold and silver, through this year and beyond.

The Vice President of Exploration for the company is John Kowalchuk, BSc, PGeo, a professional geoscientist and a member of the Association of Professional Engineers and Geoscientists of British Columbia. Kowalchuk is "very well versed in the exploration game," notes Coombes. He has over 34 years of geological experience developing and managing exploration projects and has been a director and officer of several junior mining companies.

Kowalchuk has also been involved in the discovery and staking of several significant deposits, including the Kerr porphyry copper–gold deposit, the Howard's Pass stratiform zinc–lead deposit, the silver deposit in Driftpile Creek and the Cleo Tungsten property in northern Canada. "John was credited, with Ben Ainsworth, for the discovery of the Howard's Pass deposit," says Coombes.

Benjamin Ainsworth, MA, PEng, is a Director on Columbia Yukon's board. A Senior Geologist and mining consultant who has been involved in the mining industry for over 35 years, Ainsworth graduated in 1963 with an honours degree in Geology from Oxford University in England. He joined Placer Development in 1965 and held positions of Senior Geologist, Chief Geochemist, Exploration Manager–Eastern Canada, Exploration Manager–Chile, and President–Placer Chile, South America.

Throughout the 1970s, Ainsworth was involved in the design, budgeting and implementation of exploration programs that included large and small drill programs, geophysical surveys, geological mapping, geochemical surveys, and a full range of project evaluation studies. He is also a registered Professional Engineer in the Province of B.C.

"In house, we've also got Jamie Pardy, PGeo., who used to work for the BC Geological Survey. And we were lucky to have acquired him. He looks after all of our permitting enquiries and is responsible for the drilling programs," explains Coombes. "So, John Kowalchuk looks after the exploration projects. Jamie manages the people and consultants. Then we have a GIS technician, Guillermo Funes, who looks after all our mapping, and, most recently, we have added Gary Wesa. Gary comes to us with over 35 years experience in the mining industry. Those are all of the old-time guys, and underneath them we've got a group of about eight or nine geologists that work for us in the field, managing the projects."

Columbia Yukon's Chief Financial Officer is John Morita, C.G.A. Offering his vast experience as a professional accountant, Morita is a member of the Certified General Accountants Association of BC. For the past 15 years, he has been a management consultant, advising on income tax issues, personal financial matters and has held a variety of senior positions for various public companies. Morita is also C.F.O. for International Bethlehem and Venturex Explorations.

Bruce E. Morley, LLB, BComm, completes Columbia Yukon's talented Board of Directors, with extensive experience in assisting public companies with legal and business matters. Mr. Morley is currently also a Director of Waterfront Capital Corp, International Bethlehem and Venturex Explorations. He has been a practicing lawyer and a member in good standing with the Law Society of B.C. since 1981.

"Then, we do have a Senior Advisory Board in place," says Coombes. "We've got Henry Ewanchuk with us; Mr. Ewanchuk was instrumental in the development of the Mount Polley mine for Imperial Metals. He enjoys a very well respected reputation amongst his peers. And there is Ed

Yurkowski, President of ProCon Mining & Tunnelling Ltd. Procon is one of Canada's largest underground mining and tunnelling companies. We're very fortunate and privileged to have these two individuals' wealth of experience. "So, that's it. Pretty much we try to contract all of our drilling out. We've got four drills spinning on the Storie project. And then we have two other drills that are currently drilling up in the Revelstoke area on our other projects (for Venturex Explorations and International Bethlehem).

"At the Storie deposit, right now, we've also retained Dillon Engineering. They're looking after all of the environmental assessment and permitting. So we're right now into our second year of permitting, which incidentally puts us a little ahead of the curve," says Coombes. "And doing that has also given us a good relationship with the First Nations. Understandably, they're very actively concerned about their environment. So, by doing a lot of the environmental testing upfront, it provides the data and evidence that they would like to see—that we're not damaging the environment. This works out very well for all of us.

"Throughout our exploration programs to date, we have worked very closely with the local First Nations. They're businessmen and so are we. And in treating the First Nations like fellow businessmen, you get along a lot better."

Ultimately, according to the President of Columbia Yukon, Ron Coombes, "We've successfully turned everything around and put everything in a very, very positive way here. We believe that we've got three companies now that offer significant shareholder upside potential. So, that's kind of what's happened over the last fourteen months. Now we have all of the ingredients that we believe are necessary to provide shareholder wealth."

"And an important factor to me is that we've got a very strong management team that we've assembled, led by Ron," notes Doug Mason. "Therefore, it's just the availability and timing, and the cycle of mining on a whole, that's made it something that is extremely interesting. It keeps my mind active and it keeps me 'working the weekends' so to speak," he laughs.

"Absolutely," says Coombes. "And our purpose for Columbia Yukon is to continue developing and increasing shareholder value. That's our sole purpose," he says, which means that future acquisitions are possible for the company. "But, at this moment in time right now, we're focused on the Storie moly deposit as probably the best way to build shareholder value over a short period of time."

"Of course, if something else becomes available that makes sense, we'll look at it. We may try to acquire it, either in a joint venture, or an option, or just an outright buy."

Bottom line, Coombes says, "It's just extremely important to always understand what your objective is. Our technical team: it's their objective to try to develop and expand a resource as best they possibly can. It's our job, at the helm of the company, to build shareholder wealth."

With defined and meticulous objectives, Columbia Yukon's Coombes seems to have a solid grasp on what the future may hold for the company and its Storie moly deposit. "Those who seek these types of large deposits are usually the majors, who will come knocking on your door when you're finished, or even during the process. I know we've had a number of companies approach us trying to involve themselves at a joint venture level. We're not interested in that at this particular moment in time. Basically, we haven't developed a resource large enough yet to sit down and talk intelligently about that kind of a deal. Otherwise you're simply handing off shareholder value at too early a stage.

"At the same time though, you've got to be aware that sometimes it's in your best interest to have an alliance through the process – so that your partner company can enjoy some of the upside with you – with one of the companies that may want to, at a later date, merge with you."

However, in the meantime, Columbia Yukon's path to progress is clear: "We're going to drill this thing off, significantly – if I showed you a picture of what the drill holes on the mountain side are going to look like at the Storie property; it's going to look like a thimble by the time we're finished with it. And of course then we're going to do some metallurgy tests too.

"Soon, the second year environmental study and scoping study will be complete. It's our vision that in 2008, we'll be going into a pre-feasibility/ feasibility study," concludes Coombes.

"It definitely feels good," says Mason, who is confident in the company's belief that it will imminently realize its dreams of becoming a mine.

COPPER FOX METALS INC.

Copper Fox's CEO Guillermo Salazar has practically made a career out of exploring and developing Schaft Creek in British Columbia. Salazar knows this polymetallic copper–gold–moly–silver deposit like the back of his hand. Schaft is the equivalent of his life's work coming full circle—so, who better to finally take this project into production.

As a Canadian junior natural resource mining company, Copper Fox Metals Inc. was formed by Salazar for the express purpose of acquiring an interest in the Schaft Creek, to continue what he started over thirty years ago when he first encountered the property. "I was hired by Hecla Mining in 1970, and was sent off to Schaft Creek because of my expertise, both at universities and the 'Anaconda School of Mines'—working underground, valuing ore samples, at Butte, Montana's renowned Anaconda Copper Mining Company. So, with that experience, I went to Schaft Creek to try to do some significant geology, between 1970 and 1973. At that time, the deposit was limited in size." However, Salazar's reinterpretation of the geological model of the deposit led to a massive expansion of potential resources. "I broke the limit of the deposit and expanded the resources from a maximum of 300 million tonnes to a minimum of 1,300 million tonnes," Salazar says with a smile. "And then Teck picked up the baton and added another 1000 million tonnes to the deposit and now we are at about 2.5 billion tonnes total." Copper Fox signed an option agreement in 2002 with Teck Cominco to acquire up to a 93.4% interest (direct and indirect) in Schaft and has been actively working on the site ever since.

Copper Fox's 2007 work plan has included a 15,000-metre drilling program, a scoping study of the deposit, revised pit optimization plans, metallurgical testing, and the completion of the environmental baseline study. Overall, for this year's efforts, a project budget of $17.6 million has been approved by the Board of Directors. Yet, this is just the latest in a long line of work at Schaft Creek—the bulk of which has personally been overseen by Guillermo Salazar; a man every bit as resourceful as the animal in his company's name.

"My guys jokingly tell me that they are not going to allow me to do any more condemnation drilling. Simply because, whenever I try to spot condemnation holes I find more mineralization!" he laughs. Condemnation drilling is conducted on areas around a resource where engineers want to place mine infrastructure, to make sure that they aren't going to be building on top of anything worth mining later. But, when your condemnation drilling intersects major mineralization, it can lead to resource expansion instead. This seems to be the case whenever Salazar attempts to spot condemnation holes at Schaft Creek, thus developing this substantial resource—and his reputation—during his lengthy connection with the project.

The property itself is located in the valley of Schaft Creek, in the Liard Mining Division of northwestern B.C., in an area often referred to as the 'Gold Copper Belt.' It lies 35 kilometres from NovaGold's Galore Creek and Imperial Metal's Red Chris, both copper-gold deposits, and 1,040 kilometres north of Vancouver. Comprised of 12 mineral claims covering an area of 10,269.3 hectares, Schaft Creek was first discovered in 1957 by prospector Nick Bird, who was employed by BIK Syndicate. BIK drilled three diamond drill holes to moderate depths, two of which returned copper values significant enough to encourage subsequent work. The syndicate was later reorganized into Liard Copper Mines Ltd., with Silver Standard Mines Ltd. holding a 66% interest and acting as the manager.

In 1966, ASARCO—a major producer of copper—obtained an option to explore the ground and carried out geological and induced polarization surveys. They also drilled over 3000 metres in 24 holes. The option was not maintained, despite positive drill results, and Hecla (a subsidiary of Hecla Mining Company of Wallace, Idaho) entered an option agreement with ASARCO in 1968. Hecla would earn a 75% property interest and thus commenced drilling and other exploration work.

Between 1968 and 1977, Hecla completed 34,500 metres of diamond drilling, 6,500 metres of percussion drilling, induced polarization and resistivity surveys, geological mapping, air photography, and engineering studies examining the possible development of a large, open-pit, copper–gold–moly mine. However, property work ceased in 1977, and in 1978, Hecla sold its interest to Teck Corporation (now Teck Cominco Limited). Then, in 1980, Teck commenced a program of exploration and drilling at Schaft Creek, designed to confirm and expand on Hecla's earlier work. By 1981, 60,200 metres of diamond drilling had been completed in total, and Teck undertook an engineering study to determine the feasibility of mine development.

Further data reviews were completed by Western Copper Holdings in 1988 and Teck in 1993. A combination of 230 diamond drill holes with total length of 60,200 meters, and percussion holes totaling 6,500 meters were completed prior to Copper Fox acquiring an option on the property in

2002. The option agreement enables Copper Fox Metals Inc. to acquire up to 93.4% interest (70% direct, 23.4% indirect) in the Schaft Creek deposit. "Our option with Teck is basically to acquire all of its rights," explains Salazar. "We're doing our feasibility study and one of the option clauses is that we need to generate a feasibility study with positive result. We think that we can accomplish that. We have until 2011 under the agreement with Teck to complete it, though we expect have it done much sooner than that." Meanwhile, the title to the property is in good standing before the B.C. Government until 2015.

Described as a large copper–gold–molybdenum–silver porphyry copper deposit, Shaft Creek's mineral inventory had been defined with 63,200 meters and 233 drill holes at a spacing of 76 meters, over the years before Copper Fox entered the picture. Plus, pre-1982 metallurgical testing by Teck Cominco and the Hecla Mining Co. indicated recoveries of 85% of the copper, 90% of the molybdenum, and 50% of the gold. The drill core is still stored on the property and the vast database of results gathered over the years is well preserved.

Since acquiring the project during the 2005 season, Copper Fox completed their own 15 hole drill program of 3,000 metres to confirm the integrity of the database received from Teck, as well as the repeatability of the assay results. In 2005 and 2006, Copper Fox brought the total drilled on the property to 75,415.6 metres. They also conducted their first floatation test on fresh rock from Schaft Creek. In addition, Associated Geosciences Ltd. was commissioned to prepare an NI 43-101 compliant resource calculation. The new resource, completed on June 27th and announced publicly on July 3rd, 2007, marked a significant increase in measured and indicated resources over a 2004 report on the property by Giroux and Ostensoe. The 41% increase is the result of a correction of incorrect calculations in the previous study which underestimated the copper-equivalent values of the contained gold, moly, and silver, especially in the higher-grade sections of the deposit. The new resource reported by Associated Geosciences Ltd. contains 1.393 billion tonnes, measured and indicated, at a 0.2% copper equivalent cut-off, which includes 7.7 billion pounds of copper at 0.25%; 8.1 million ounces of gold at 0.18 g/t; 584 million pounds of molybdenum at 0.019%; and 69.4 million ounces of silver at 1.55 g/t.

The new and updated resource estimate was completed by Riaan Herman under the supervision of Keith McCandlish, a Qualified Person under NI 43-101. Meanwhile, Dr. Gilles Arsenau, P.Geo, of Wardrop Engineering, Inc., also a Qualified Person, reviewed the copper equivalent formula, the mineral resource classification, and block modeling methodologies. The site's current geological model is the work of Walter Hanych, Dr. Peter Fischer, and Sheena Ewanchuk under the supervision of Copper Fox CEO Guillermo Salazar, MA, PGeol.

The updated resource actually presented the possibility for a much larger operation than previously contemplated at Schaft Creek, a site that is well-versed at expanding its limits. For the first time, the presence of a higher grade resource was noted consisting of 208.9 million tonnes grading 0.42% copper, 0.34 g/t gold, 0.031% Mo, and 1.52 g/t silver with a conservative and recoverable copper equivalent of 0.66%.

Naturally, these impressive copper numbers speak for themselves. Nevertheless, despite the company's name, Salazar says that the accompanying moly and precious metal values didn't come as a surprise to him. "This deposit was always known as a copper–gold– molybdenum–silver deposit." Salazar continues, "There were suspicions that there was another element in there as well—rhenium. Obtained as a byproduct of molybdenum refinement, rhenium is one of the ten most expensive metals on Earth. And without rhenium in our propellers and jet engines, they just don't fly. There's no unleaded gas without rhenium (rhenium is now used as a catalyst in unleaded gasoline production). But, we have to do a lot of work on that before we can prove the economic parameters of it."

As a contender for one of the largest copper deposits undeveloped in Canada today, Copper Fox is now evaluating the feasibility of developing this open-pit mine with a minimum capacity of 100,000 tonnes per day and a mine life of 35 years. The company has also entered the Environmental Assessment process required by the British Columbia and Canadian governments for project permitting. In fact, Schaft Creek's environmental and socio-economic baseline studies were started back in October 2005 and are moving forward. The studies included wildlife (moose, goat, and bird studies), water quality, aquatic biology, fisheries, hydrology, meteorology, archaeology, and metal leaching and acid rock drainage (ML/ARD). As an ongoing process, the scope of work for the 2007 environmental and socio-economic studies was increased with a view to fulfilling requirements of both the governmental assessment process and specific studies requested by group, the Tahltan Nation. These include: socio-economic impact, traditional knowledge, country foods, wetlands, hydrogeology, soils, ecosystem mapping, vegetation, archaeology, human health, fisheries and aquatics, wildlife, hydrology and ML/ARD. The baseline studies from 2005 through 2007 will form the basis of the Schaft Creek EA application. The anticipated completion date is the fourth quarter of 2008.

In the meantime, bench and bulk flotation tests were conducted, with results expected by the fall of 2007. Recoveries of the four main metals that can be expected in a typical flotation mill are being defined with one and two tonne tests, now that laboratory scale recovery testing has been concluded. Copper Fox's metallurgical consultants are working on their reports. "So, what we're doing now is speeding up the process for the feasibility study," says Salazar. "The scoping study will be completed first. It will include the results of our new resource calculation, pit optimization plan, bulk metallurgical results and will give us a first look at the economic feasibility of the deposit using a small proportion of the vast resources defined at Schaft Creek to date. This preliminary pit will be designed to fast track the development of the project and the re-payment of the required capital expenditures."

2007 is proving to be a busy time for Copper Fox, Salazar says, "We have between 55 to 60 people at the Schaft Creek camp right now. Probably about 40 are from the north and the other 15 to 20 are either from southern British Columbia or Alberta." A 1,000 metre long airstrip lies adjacent to the camp. Four drills were on site during the summer 2007 season. Two drills have been performing geotechnical work, and a third drill tested the tailings dams. On completion of a pit optimization study—being designed by Moose Mountain Engineering Services Ltd. under the guidance of Samuel Engineering Inc.—the fourth drill is set to proceed with the work designed to test the pit's slope stability and the possible extensions to the mineral deposits. The upcoming scoping study will evaluate the possibility of operating the mine at a proposed 100,000 tonnes-per-day capacity, based on Copper Fox's recent work. The scoping study will also take into account road infrastructure at Schaft Creek.

Currently a 'fly-in, fly-out' scenario, Copper Fox is looking into the possibility of constructing a road to link up with Novagold's proposed road access which goes from Highway 37 to its Galore Creek project. McElhanney Surveys has been hired to do a preliminary road layout and survey of this scenario. This work is once again under the supervision of Samuel Engineering who are mandated to prepare the Schaft Creek feasibility study. Additionally, B.C. Hydro has been contracted to provide a preliminary evaluation of the power situation at the site. Power availability and the type of power are big issues for Copper Fox.

Considering the overall timeline for continuing work at Schaft Creek, Salazar lists the next steps for the project: "The milestones have not changed. We still expect to have our feasibility study finished by the end of December, 2008. And our scoping study will be finished sometime in

December, too. As mentioned, we are still running metallurgical testing and we have those four drills on the property doing all kinds of geotechnical work to make sure that we will be able to have—at least—the mining parameters required for developing Schaft Creek into an operation of a minimum 65,000 to 70,000 tonnes a day."

Predicting a 30 to 35 year mine-life for an operation at Schaft Creek, Guillermo Salazar is convinced about its bountiful and productive future—and the fact that the mine may far outlive Salazar's own concerns. "Well, the lifespan of the mine is something that my great-grandchildren are going to have to worry about, you know," he says.

If asked whether the location of the company's headquarters, in Calgary, Alberta, is a strategic move, Salazar admits, "Yes, it's strategic—my grandchildren live there. "Really, where the company's presence is going to be will depend on how the company grows. 25% of the shares are owned by Albertans but the asset is in British Columbia. Copper Fox will have a stronger presence in northern and southern B.C. – we need one because of the size of the asset – as the company grows. There is no doubt about it."

Since January 1980, Guillermo Salazar has been Calgary-based as a self-employed consultant. From 1995 to 2000, he was Chairman of the Board, President and a Director of Volcanic Metals Exploration Inc.—formerly OroGrande Resources Inc.—a publicly-traded company on the TSX-V. Salazar received an Engineering Degree in Mining at the the National University of Engineering, a prestigious engineering and science university in Lima, Peru, and an MA in economic geology from Harvard University in Cambridge, Massachusetts. He was granted the designation PEng and PGeol by APEGBC in June 1976. Now in the twilight era of his career, at age sixty-three, Salazar acknowledges that he isn't intent on hanging up his mining hat quite yet, but he is considering it. "I do believe in Teck's mentality that at the age of sixty-five, a fellow should be thinking of retiring. And it doesn't mean that he does, but he should be thinking about it." In which case, "that would definitely mean there would need to be a young replacement ready to take the ball and run with it." With things coming to fruition at Schaft Creek—a property he knows so well—Salazar can imagine the perfect retirement present. "It would be a beautiful situation to break sod at Schaft Creek, and say: 'Well, it's time for me to go,'" chuckles Salazar.

While moving steadily to the anticipated climax of creating a mine, Copper Fox has been working equally hard at creating a great team. "We've been working, and working hard on the team," Salazar says. "Getting this team together was one of the biggest accomplishments that we've had."

That team includes J. Michael Smith, Executive Vice President and Director; Murray J. Hunter, CFO and Vice President; C, Frank Agar, Special Advisor and Director; Darren B. Fach, Assistant Secretary; Elmer B. Stewart, Director and Chairman of the Board; and Cam Grundstrom, Vice President Operations.

Until his retirement in November 1999, Michael Smith was part of the senior management team with RBC Royal Bank in Edmonton Alberta. During his banking career, he also held management positions in Calgary, Alberta; Montreal, Quebec; and in New York City. Since retiring, Smith has acted part-time as a consultant to businesses including a private Calgary-based mineral exploration company and Copper Fox. In addition, he is a Director and Treasurer of a registered public charity.

Murray Hunter, CA, is currently owner of Murray Hunter Professional Corporation and has been a member of the Institute of Chartered Accountants in Canada since 1975. His financial experience includes positions as an Investment Advisor for BMO Nesbitt Burns from 2000 to 2003, as a Financial Consultant for CIBC Woody Gundy from 1996 to 2000, and as a Managing Partner for KPMG from 1990 to 1995.

Frank Agar, BSc, PEng, has been a self-employed consultant in Calgary since May 1988. Over the years he has served as a director of many publicly-traded companies, including Shear Minerals Ltd., Stratabound Minerals Corp., Three Sisters Resorts Ltd., and Destination Resorts Inc. Agar received his Bachelor of Science degree in geological engineering from the University of Saskatchewan and was granted the designation PEng by APEGGA in May 1967.

Darren Fach is currently a partner with the law firm of McLeod & Company LLP, where he has practiced since 1992. He was admitted to the Law Society of Alberta in 1990. Fach has also acted as corporate secretary of Elarra Ltd. from March to September 2002 and as secretary of Sheffield Resources Ltd. from December 1999 to October 2004. He continues to fulfill the role of secretary to Eagle Plains Resources Ltd.; Superior Canadian Resources; and Telkwa Gold Corporation.

Elmer Stewart, BSc, MSc, PGeol, is President of Alhambra Resources that owns and operated a 15,000+ ounce gold producer in Europe. Stewart has been a director of several junior mining companies, received his Bachelor's and Master's Degrees from Acadia University and was granted the designation of P.Geol. by Apegga in November, 1983. Previously, he served as President of Eurasia Gold Corp. and Kelman Technologies Inc., and as Executive Vice-President and Corporate Secretary of Capilano International

Inc. Stewart has also been a director of a multitude of public companies. He received both his Bachelor and Mater of Science degrees in geology from Acadia University. He was granted the designation P.Geol. by APEGGA in November, 1983.

"I'm very proud to be involved with them," Salazar says of his team, "and their contributions are very special to the company. Murray Hunter as a C.F.O. is our conscience, if you will. He makes sure that we do things right. Michael Smith, he's always implementing his experiences that he learned while working for a very large bank, with over 3,000 employees. And Frank Agar, he is another fellow that has been around for a long time. He was the one that put the northernmost mine in Canada into production. While he was President of Mineral Resources International Ltd. and Nanisivik Mines Ltd., Agar financed, developed, and operated the Nanisivik zinc, lead, and silver mine on Baffin Island. And that alone gives him an insight into something very different from what the rest of us have," Salazar notes.

"And now there's Cam Grundstrom, who was the most recent one to join us. Cam and I have been friends for a long time. We worked together back in the late 1970s and early 1980s. Then he went off and gained a whole lot of experience in large porphyry copper deposits with very difficult conditions." A mining engineer with extensive operations and development experience, Grundstrom held key management positions in large mining projects in Canada, U.S.A., and overseas during his 26 year career, including assignments in world-class underground and open-pit mines. Most recently, Grundstrom worked for BHP Billiton Diamonds Inc. where he was Mining Coordinator at the Ekati Diamond Mine, Northwest Territories—Canada's first diamond mine.

"So, Cam's our Vice President Operations. He will have a major role in the development of Schaft Creek, and he's the designated successor for my position, actually. Though if he doesn't want to take the job, there are three others on the board of directors that would also be qualified and able to take on the role." Obviously, Salazar's shoes would be tough ones to fill. Not only does he have first-hand knowledge of the nuances of the property, since first working on it over three decades ago, but he has also been deftly applying his experiences in Peru to Copper Fox's consultations with the local First Nations people, the Tahltan.

"The Tahltan Central Council is very much pro-business. We're endeavoring to develop a very good relationship with the Tahltan people and we are always open to ways we can benefit each other," says Salazar.

"I come from Peru originally, and the Peruvian mines are fairly remote. You depend on your neighbour, and your neighbours depend on you. So, from that perspective, I guess it's something that has grown on me and on us, that we need to have good relations with our neighbours." The forging of a mutually-beneficial relationship has resulted in a communication agreement between Copper Fox and the Tahltan Nation. The agreement relates to sharing information on a regular basis through engaging community members at scheduled public meetings and eliciting feedback every step of the way. Working collaboratively, Copper Fox is also entering into a funding agreement with the Tahltan Heritage Resources Environmental Team (THREAT), so that this organization can similarly participate in the review process as the company earns the respect of the Tahltan and other communities.

As of May 2007, a Memorandum of Understanding (MOU) was completed between Copper Fox's CEO and the Tahltan Nation Development Corporation's President. As the economic arm of the Tahltan Nation, the TNDC signed the agreement that sets out a joint understanding and intention to cooperate in carrying out the work at the Schaft Creek mine project. The MOU defines the scope of work, program commitments, and cooperation that Copper Fox will follow, and recognizes the TNDC as a "preferred contractor."

Therefore, with a budding relationship in place between Copper Fox and Schaft Creek's neighbouring community, as well as a professional executive team, and a promising polymetallic property, it definitely seems like Guillermo Salazar has all his ducks in a row. Furthermore, with the driving demand associated with 'molymania' and global industrialization, Copper Fox appears to be perfectly positioned. "When you're looking at a deposit that has 584 million pounds of molybdenum; that is a big deposit," concludes Salazar. "We definitely think of it as something that would produce about 20,000 tonnes of molybdenite concentrate, in a year. That is a significant production."

Nevertheless, Salazar notes, "The project is sensitive to metal prices, of course, and the price of copper is the number one; it's the one that affects us the most. The price of moly and gold are next. And if moly stays at a minimum price of $20 a pound, it's going to be a moneymaker. And, you know, here we are talking about three metals, but we're getting numbers for silver at Schaft Creek that indicate we'll be producing about one million ounces of silver, a year. When you multiply that by, say, even 20 years—that's a lot of money." His price predictions suggest he is optimistic that metal prices across the board should remain strong throughout Schaft

Creek's development. "Copper is staying above $2.50," he suggests. "Gold is staying above $500; molybdenum is staying above $15; silver is staying above $7.50; and don't ask me about rhenium because the sky seems to be the limit. These are conservative long-term projections of metals prices."

Recognizing the value of Schaft Creek's potential, Guillermo Salazar says that they are consciously centering the company's attention solely on this one property, and there aren't any plans in motion to acquire other properties in B.C., or elsewhere. "Right now, we are a lean and focused organization," he smiles. "We have developed into a company with so much in-house expertise that, although my expertise may be developing a feasibility study; two or three members of our internal team are more than capable to run a large mining company. So, at any time they can step in and I can go golfing..." Salazar continues, "Then, somewhere in the near future, hopefully we will be as successful as we think we will be. Then, we will be looking to apply our success to other projects."

EAGLE PLAINS RESOURCES LTD.

Eagle Plains Resources Ltd. collects properties like stamps. The company has a double-digit tally in their impressive inventory of properties. However, President and CEO Tim Termuende isn't merely subscribing to the "safety in numbers" school of thought. In fact, work is ongoing on about half of Eagle Plains' prospects. This is all part of Eagle Plains' holistic strategy to maximize shareholder value and develop these early-stage exploration projects to the next level.

"Yes, we're working on many of the properties in 2007; we'll likely have fifteen different programs," says Termuende. "That's why I'm in such a hectic travel mode right now. We're just bouncing from project to project. And, we've got roughly 50 people out in the field for us at the present time."

Termuende admits his drive to add value to the company's shares is not totally altruistic. "I'm one of the biggest shareholders," he laughs.

"It's a good place to be," Termuende says of the current supportive environment for resource development in western Canada, "and part of our success right now is because we stuck it out in the dark years. You know, when everyone else was leaving the business and going into dotcoms or going to work on projects in South America and Asia; we stuck around and chose from a huge smorgasbord of projects that people left behind in the exodus."

He's not kidding about the smorgasbord. There are approximately thirty-five properties packed onto Eagle Plains' expansive plate and they run the gamut, from base to precious metals. Eagle Plains holds gold, silver, uranium, copper, molybdenum, zinc, and rare earth mineral projects; two of which have NI 43-101 compliant Inferred Resources. The Blende property contains 35 million ounces of silver, 1.2 billion pounds of lead, and 1.3 billion pounds of zinc and the Sphinx property contains 48 million pounds of molybdenum.

"We managed to pick up a few properties that actually have deposits on them, and we basically got them for the cost of staking," says Termuende. "I think of the properties we picked up from 1999 to 2001, which were the darkest days of the nuclear winter we went through here; seven of those were deposits. We calculated what money had been spent on them over the years—about $38 million in exploration expenditures in today's dollars. Yet, we got all seven for about $100,000 in acquisition costs."

Their diverse cornucopia is predominantly situated in British Columbia and the Yukon Territory, along with a couple of properties in the Northwest Territories and Saskatchewan. "We're not interested in going to Alaska or to Peru or anywhere else. Through the tough years, we set up a network with prospectors, suppliers and drilling companies. So, we've got very good relationships established here in Western Canada. We're very familiar with the geology, we're very familiar with the government people, and with the province's established infrastructure, we just find it's a good place to be, says Vancouver-born Termuende, who now works from his Cranbrook, B.C. headquarters. "A lot of people are now turning to British Columbia as an exploration hot spot, while we've been here for a long time, so we're established."

Unquestionably, in B.C., Eagle Plains is very well established. Its alphabet soup of properties include the Acacia, Black Diamond, Bohan, Bootleg, Coyote Creek, Elsiar (LCR), North Findlay, South Findlay, Greenland Creek, Hall Lake (Cretin), Hot Punch, Ice River, Iron Range, K9, Kalum, Kokanee Creek, Sphinx, Titan, and Wildhorse. In the Yukon, Eagle Plains holds the Blende, Dragon, Hit, IronMike, McQuesten, Pelly Mountain, Rusty Springs, and Sprogge. In the Northwest Territories, there are the Gayna River and Mackenzie Valley Zinc projects and in Saskatchewan, the Karin Lake, Eagle Lake and Kulyk Lake projects.

Eagle Plains has been known to 'spin-off' a property into a brand new corporate entity, in order to spotlight the project's value, and create value with new shares, as well as making a project available for outright acquisition by a producing company. "The best example of this model is our Copper Canyon property, which is now the flagship property of Copper Canyon Resources Ltd.. Until last June, that was just another one of our thirty-five projects."

When Tim Termuende pays tribute to Western Canada as a superior place to develop mining projects, he speaks with comprehensive knowledge. In addition, he has international experience as a professional geologist who has worked on exploration projects throughout North, Central, and South America, and inspected mineral deposits in China and the former Soviet Union.

"Oh yeah, I've seen a lot of nightmares offshore. You know, tenure, political and Indigenous issues and all kinds of crazy things. We just feel British Columbia provides a very stable platform to carry out what is already a risky business. Exploration is inherently risky, so if you can mitigate your risk in any way you can, it just makes more sense," says Termuende, who has over thirty years experience in the mineral exploration industry.

Despite the breadth of his international experience, the work abroad never managed to distract him from his bold plans to build a new junior exploration company of his own. "The international work kept food on the table while my father and I were building Eagle Plains. I was doing work as a consultant for other companies that were working overseas. And again, most of the travel and most of the overseas work I did was when B. C. was dead. So I could bring home a paycheck, then go and stake some claims in B.C.," smiles Termuende. "That's essentially how that worldly experience paid off. It was actually in the background, helping to build Eagle Plains."

Eagle Plains was founded by father and son in 1992, and perhaps one day a grandson will be involved, too. "I started the company with my dad, Robert (Bob), and we went public in 1995," says Termuende. "In fact, I graduated from the same program at the University of British Columbia thirty years after he did," earning a degree in Geological Sciences in 1987. "My father's a geologist, so am I, and it looks like my oldest son might be going into geology, too. So, it's a family passion, or obsession – whatever you want to call it." Now in his mid-forties, Termuende adds, "I'm one of the younger CEOs at any of the industry conventions. I always have been. I mean, I was always the youngest in this business. My first summer of work was in northern Saskatchewan when I was twelve years old. And now my oldest kid's had his first full summer. He's fifteen and has just finished a full summer working up in our Mackenzie Valley camp. So, we've got three generations at work here."

The Termuende family's generations of knowledge about Canadian mineral properties has recently led Eagle Plains a little further east. "We've detracted a little bit from our general motto of 'Western Canada only' by picking up a couple of uranium projects in Saskatchewan. I thought we might have been too late to get in on uranium, but I talked to my father—who's now retired—and who had worked extensively in the Athabasca basin in the 1960s. A couple of projects really stood out in Bob's memory and he suggested Eagle Plains look into them. "And, sure enough," Tim says "they were still open ground." Eagle Plains quickly picked them up and soon after had formed three joint ventures to explore them.

Additionally, the Eagle Plains CEO recently returned from a trip to Canada's Northwest Territories. "I was working on our Mackenzie Valley

Zinc project. We've got a crew of 25 people up there right now; we're in an alliance with Teck Cominco on that one. So, that was a great deal, being able to attract those guys. They bought a bunch of shares and did a private placement at a pretty steep premium to the market price and we're spending that money up there now. I think the relationship with Teck Cominco will stand. Not many junior companies have exposure to zinc. So, now they're trying to leverage their zinc exploration through companies like Eagle Plains."

Termuende then admits that he is a faithful supporter of base metals, despite owning several precious metals properties as well. "It's a dangerous allegiance," he notes. "The newsletter guys and the analysts don't like it. They want everybody to be after gold, but my heart's more in base metals. I honestly think that's where our home-run is going to come from. Copper Canyon (a copper-gold project), for example, was a nice base-hit, maybe a double. And the molybdenum is something we've found along the way. We just sold a gypsum deposit and we weren't looking for gypsum either—we just found it while we were looking for zinc," laughs Termuende. "We sold that to a major gypsum entity. So, we're basically always looking for lead–zinc and we find some other minerals along the way.

"Actually, I started with Cominco when I graduated from UBC, and I trained as a lead–zinc exploration geologist. Lead-zinc are not really sexy metals to be after and they don't get the headlines that a gold project does, but I still think, in the end, a couple of years down the road, it's going to be base metals that really make Eagle Plains successful. Part of that is, like I said, there are very few juniors that look for base metals. And that was part of the big story with us and Teck Cominco; they recognized that with us. Now we're in a strategic alliance with Teck Cominco on the Mackenzie Valley Zinc project, and that's hard to do—those guys don't often jump in with juniors. They do a lot of research and, you know, it was a long journey to finally get to the point where we signed that agreement. But, they wanted to make sure they knew what they were buying from us, too—and they're very happy. "

Of Eagle Plains other properties, the already NI 43-101 compliant Blende and Sphinx properties top the pile. "I really like the Blende property, partly because it's more advanced. It's got a resource on it right now. And we're building a resource on the Sphinx moly project as well," says Termuende. "Though we're not at the feasibility stage yet on these things – we're not even at the scoping study stage. Basically, we're still in the exploration stage and we need to convince ourselves and others that there's enough there and that the grade is high enough. We need to be certain that these things have a good chance of going into production.

"Most of our projects are grass-roots. We're often the first guys to work on them, and initially we have no idea what's there. It's usually a long, arduous, expensive process to bring a property to the point where you want to make a decision whether to put it into production or not. The Sphinx and the Blende are getting much closer to that point."

Eagle Plains controls a 100% interest in the Sphinx and the adjoining Jodi properties. The collection of claims consists of a total area of 2,500 hectares, and is located 60 kilometres west of Kimberley, B.C. The large land package includes all significant claims in the area, including lands suitable for mine infrastructure and tailings areas. Moreover, it is road-accessible and has been logged extensively. The properties are situated alongside a high-voltage hydro-electric line and rail facilities are located at Marysville, only 60 kilometres east of the property. Aboriginal claims and permitting issues haven't arisen at this stage, although Termuende says, "We're doing the field work, we're keeping the relations friendly with everybody, so that when someone else comes along to put it into production; they have no skeletons they have to deal with."

Geologically, Sphinx boasts molybdenum and associated tungsten mineralization, occurring as quartz-pyrite stockwork veins that are hosted by both sedimentary and intrusive rocks. According to studies, the chemical alteration of the rock units suggests the presence of a substantial porphyry-style mineralizing system.

Historically, the project area was first identified by Cominco Ltd. in 1978, which carried out surface work and limited diamond drilling until 1984. Cominco completed a soil geochemical survey which resulted in the delineation of a 1,700-metre by 500-metre tungsten–moly anomaly. Four to six drill holes were completed, but no results were ever released. Consequently, in 1997, Barkhor Resources drilled ten holes into the soil anomaly and encountered significant mineralization over a 1,000-metre by 300-metre area. Results from only one hole were released, but a private consultant reported that "typical drill intersections are averaging 0.03%–0.038% Mo over lengths ranging from 90 to 230 metres."

In 2004, Eagle Plains completed an airborne geophysical survey that outlined a large intrusive feature associated with the previously-defined soil anomaly. All the drill core from the historical programs by other operators has since been secured and examined by the company. The data was compiled and reported in May 2005. Then in July 2005, Eagle Plains completed a fourteen hole, 3000-metre diamond drilling program, and reported widespread mineralization in most holes on the Sphinx moly project. Nearly all holes intersected significant molybdenum mineralization over a broad area, with the mineralized zone open to depth and along strike in one direction.

During the winter of 2005–2006, Eagle Plains commissioned Barry Price, P.Geo., to compile an NI 43-101 compliant technical report calculating a resource for the Sphinx. In his report, Price outlined an Inferred Resource of 62,005,615 tonnes grading 0.035% Mo, using a cut-off grade of 0.01% Mo for a contained metal calculation of 47,844,630 pounds of molybdenum. Following on the heels of the 43-101 report, two more drill programs were carried out. Eagle Plains drilled 1700 metres in the fall of 2006 and another 2350 metres in the spring of 2007. The current program is following up on those results. The results of all three of these programs are expected to be used to update the current inferred resources calculation, to further expand this substantial moly asset.

But as Termuende explains, Eagle Plains "picked up the property before molybdenum was a hot commodity. It was just that the opportunity came up. Those were the dark days, so it was a fairly soft deal and we decided to take a shot at it, as it was a property that had been worked in the past and as a result, had a lot of valuable data with it. But, it was another one of those things that had a bunch of drill core sitting there that no one had looked at in ten or fifteen years. So, we re-assayed a lot of it. We put all the data together into a database and just started working it.

"My ultimate goal for the Sphinx is still to find a partner that wants to come in and get involved with us on it. But, until then, we will keep doing work on it, as we have to convince people that this thing has a good chance to be an economic deposit. And we're convinced it does, or we wouldn't be spending the money we're spending on it," he says. "But, ideally and ultimately, we would like to have some big moly company come and say, 'hey, we'd like to be a part of that project.' I think that'll happen. It's just going to take more work on our part, more drilling, and more money to go into it."

The Blende is a slightly different story. "Some of the projects we picked up were worked fairly extensively in the 1970s. The Blende, for example, had $8 million spent on it, right up until 1994." Billiton Metals Canada Ltd. carried out diamond drilling in 1991 and delineated two zones. "Yet, later, in 1999, those claims became open, so we swooped in there and grabbed them," explains the Eagle Plains' President.

A silver, lead, zinc prospect, the Blende is known as a carbonate-hosted deposit, secured by 100 claims, located approximately 65 kilometres northeast of Keno in central Yukon Territory. Based on exploration work to the end of 1991, an NI 43-101 compliant resource was reported as 19.6 million tonnes grading 55.9 g/t silver and 5.85% lead-zinc containing an Inferred Resource of 2.48 billion pounds of lead–zinc and 35 million ounces of silver, located in the two zones.

On surface, the deposit is outlined by an open-ended 4.8 kilometre-long soil anomaly which contains values up to 10,000 ppm zinc. The Blende is exceptionally responsive to most geophysical methods including IP, VLF, and Max-Min EM, due to the inert nature of the host-rock, dolomite. It has been established that the deposit is non-acid generating and could be mined by open-pit methods, with a stripping ratio of 2.1:1. Plus, preliminary metallurgical studies indicated no significant concentrations of deleterious elements. "Currently, we're going through all the old data—a lot of it was never digitized. There weren't many computers around when they were doing much of that work so we're taking all this data that really has a value of $8 million or, at today's prices, more like $20 million. That's value that doesn't go away.

"And that's what we've managed to accomplish," Termuende says, "securing all those projects and accumulating all that data, which is as valuable now as it was back then. So, we've added to that a GIS database, we're merging it with new data that we're creating constantly right now, and we're just building a picture on a number of these projects."

Understandably, with lots of projects comes lots of work: Eagle Plains has at least ten drill programs scheduled for 2007 alone. "We actually bought a drill company last year to make sure we have a steady supply of drills. We did that to get around this drilling problem that we saw coming, and many exploration companies are experiencing now. It's so extremely frustrating, and that's what we went through last year. But we solved that problem and now we've got these two drills that are dedicated exclusively to our projects this year," he explains. "Five years ago, people had no money, and drilling companies would basically fall at your feet to get any kind of work, and at any kind of price. Now, all these junior companies have lots of money and they can't get drills, they can't get people, they can't get labs to respond. Anyway, we saw that coming and took pre-emptive strikes and now it's paying off big time for us," Termuende says.

At the end of August 2007, Eagle Plains commenced a diamond drilling program on their 100%-owned Findlay base-metal project, 35 kilometres northwest of Kimberley, B.C. Drilling activity follows up on extensive exploration work completed by Eagle Plains and past partners Rio Algom, Kennecott, and Billiton Metals during the 1990s. The project is under the supervision of David Pighin, P.Geo. Additionally, drilling activity has been completed on their 'Bronco' discovery, in the Mackenzie Mountains, near the Yukon/NWT border. Five holes were completed for a total of 807 metres. Also, 259 metres of drilling activity has recently been completed on Eagle Plains' 100%-owned Ice River project, located in southern B.C., which has lead, zinc, copper, iron, silver, and gold values.

Ultimately, Termuende believes that this recent work—on the Sphinx and Blende, and Eagle Plains' various polymetallic projects—will result in the biggest pay-off. They've clearly kept their options open and as Termuende says, "there's nothing wrong with diversity. We've got at least some exposure to just about every metal under the sun, and, you know, they all become the flavour of the month every now and then—and we'll be there."

Eagle Plains is as well placed as it is to take advantage of the growing demand for many different metals because they managed to stick it out through the lean years. Termuende remembers those lean years. "We were broke. At one point, we had $30,000 in the bank, as a public company. And now we've got $10 million in the bank, and that's only over about four or five years. It's happened quickly. But, there were a lot of times when I was wondering whether or not to jump—everyone else was leaving, but luckily we had a couple of directors who had some money from other businesses, and they kept buying private placements at 5 to 10 cents. We used that money to acquire projects; we were constantly acquiring."

Tim Termuende nods and looks around his office in the small town of Cranbrook. "The head office of the company was basically in my basement for ten years," he says. "Our overhead was virtually non-existent. And that's one of the reasons we managed to survive and even thrive through that period too."

He concludes, "I think there's a big gut-check going on right now, across all the equity markets. But, I think it's a healthy correction, too, and I also think the mining business has at least ten good years in it. It's just simple math in my mind on that, because it had ten really bad years. You can't have ten bad years, without ten good years, because there's supply and demand. In the bad years, people weren't looking for new metal sources, but they were still consuming. We're seeing unprecedented development in

China and India right now, against a back-drop of dwindling metal supply. We're seeing the formation of the perfect storm.

"There are a lot of mines closing down right now, yet there's no replacement for them. And so we're running out of zinc right now, we're running out of moly. What you're going to see are big, big gaps in supply and demand, and that's going to cause even more frenzied price escalations, which is going to cause more exploration, which is going to lead to more discoveries. Nickel, lead, zinc you name it, they're all at record highs right now. But, I think they're still going to go higher because you can't discover and start producing these metals in a short time period. It's seven to ten years to get to production, no matter what. And I think we're just going to run into more and more deficits for metals. We've had two good mining years, and I think we're going to have another eight, at least," says Termuende.

As with the company's prior success with Copper Canyon, Termuende is optimistic that in these better years ahead, there could be further spin-offs to the benefit of Eagle Plains' shareholders. "Well, Copper Canyon is now worth $40 million, and basically we gave our shareholders a 1-for-1 share, so if you had 1,000 Eagle Plains shares when we did the spin-out, you ended up with 1,000 Eagle Plains shares and 1,000 Copper Canyon shares too. Everybody's very happy. And we're hoping we'll be able to do that with other Eagle Plains projects. Maybe it'll be the Sphinx, maybe it'll be one of the uranium projects, who knows? Once one of these projects stands out from the rest of the "herd", we plan on separating and isolating it.

"By separating them, it's kind of like splitting the atom—we just release value that was always there, but was locked up. I think if we hadn't separated Copper Canyon from Eagle Plains, we'd probably still be trading at relatively low levels, and there would be no other value realized from that. So, it was an exercise in creating value from what we already had.

"You know, that's what you have to do—look out for the shareholder. They are number one."

Along with spinning-off companies, Termuende is evidently aware of the potential benefits of joint venture opportunities. "With all those projects, we try to joint venture them as much as possible," he says. "Of course, it's a risky business. You've got to kiss a lot of frogs and you've got to be prepared to walk away with nothing on some of these programs. So, the more we can spread the risk out and the more we can use other people's money, that's the whole business of risk management. Because, you're looking for something that's hard to find and, the odds are, you won't always find it. In which case, you've got to just try and manage your risk as best you can."

The company is considered a project generator and consistently invites joint venture participation to expedite the development of its wide inventory of properties. This strategy reduces shareholder exposure to the risk of exploration and enhances exposure to a discovery. To date, Eagle Plains has controlled projects with third parties such as Teck Cominco, Alexco Resource Corp., Wellstar Energy Corp., Blue Sky Uranium Corp., Blind Creek Resources, and Golden Cariboo Resources Inc. These agreements expose Eagle Plains to over $17 million in exploration expenditures over the next five years.

In recent years, Eagle Plains has completed option agreements with Billiton Metals, Rio Algom Exploration, NovaGold Resources, Kennecott Exploration, Viceroy Resource Corp., and numerous other junior exploration companies, resulting in over 38,000 metres of drilling and over $16 million in exploration spending on its projects since 1998.

"With our in-house staff we are able to technically assist our joint venture partners. And if the partners we've attracted have a very good exploration team, we let them do the work, and if they don't, we insist that we manage the programs with their money" says Termuende.

Termuende has as much confidence in the Eagle Plains board as he has in the company's exploration team. The committed board includes Glen J. Diduck, CFO; Charles C. Downie, PGeo, Vice President Exploration; Darren Fach, LLB, Corporate Secretary; David Johnston, BASC, MASC, Director; and Ron Netolitzky, MSc, Director. "Chuck Downie and I worked together in the 1980s with Cominco," notes Termuende. "He's worked in all of Cominco's major mines—Pine Point, Snip, Polaris, Sullivan. Now, he's our VP of Exploration. He's controlling and directing all our exploration activities on our projects."

Termuende highlights, "We've actually got a schedule that we put together, where we're moving our people. We can't work on fifteen projects at once, but we're working on fifteen projects over five months. So, we're two weeks here, two weeks there. We've been able to stay pretty close to our schedule."

Accordingly, as the exploration proceeds—meeting Eagle Plains' targets and the expectations of joint venture partners—the future development of a mine inevitably rolls into view. "Sure, a lot of companies think, 'we're going to find something and mine it.' All of a sudden, they start hiring miners and engineers and buying Caterpillar trucks and bulldozers. But, that's not our strength. Our forte is exploration, and we're very aware of that," he says matter-of-factly.

"I think sooner or later our spin-off company Copper Canyon will get sold 100%, just outright, and I'd be happy to see it go—because it'll be a benefit to the shareholders and will allow us to focus on what we do best. The same thing may happen with Sphinx. We have no illusions that we're ever going to design a mine at Sphinx. We're going to let somebody who knows how to do that, do that. We're paving the way, but again, we have no intention to mine this thing. That's not our job, that's not our place in the food chain," says the Eagle Plains CEO.

He continues, "Really, it's tough to acquire projects now. For the most part, we're out of seeking new acquisitions right now; we're more into aggressively working our projects. We've got our drills and technical teams. We've got partners who are providing much of the funding.

"We've got what everybody wants now: We've got projects, we've got people, and we've got two drills," laughs Termuende. "Now it's a matter of how we best capitalize on that."

ERDENE GOLD INC.

Few of us go half a world away to go to work. Peter Akerley, President and CEO of Erdene Gold Inc., lives and works in Halifax, Nova Scotia, but he leaves his wife and children several times each year, to travel to the area which could well make his company one of the world's most profitable and famous producers of the valuable mineral molybdenum: Mongolia. "From the Halifax airport to the capital Ulaanbaatar takes 28 hours, and then from there to our major property, Zuun Mod, you're looking at a two-day drive, although you can cut that down by a day if you fly to one of the local air strips." The apparent remoteness is somewhat of an illusion as just 200 kilometres from the project is the Chinese border across which you find a massive industrial complex stretching across northern China with insatiable demands for metals to fuel their enormous growth. Steel mills generating stainless steel and high strength steel products and demanding large amounts of molybdenum continue to grow. China has the highest rate of consumption growth of molybdenum in the world. Some experts say China will begin importing large amounts of molybdenum by 2010.

Mongolia. There is a name you don't see in the papers every day—or its capital city, correctly spelled Ulaanbaator. On a map, you see gigantic Russia to the north, and enormous China to the south (the first and third geographically-largest nations on earth, with Canada coming in at number two). Erdene's corporate literature explains why this Canadian company finds the faraway, exotic nation extremely attractive:

Mongolia is highly prospective for the discovery of world-class mineral deposits. Major ones can be found in and around the country. These deposits occur in tectonic terrains that typically host numerous large deposits. However, Mongolia remains largely under-explored. Erdene's properties are well situated within these known terrains and include areas where significant mineralization has been intersected during Erdene's drilling programs.

The above refers not only to Erdene's copper and gold interests; the firm has several other minerals on the radar there, as well. Perhaps most exciting of all is their Zuun Mod molybdenum project (Zuun Mod means "100 Trees") located in south-western Mongolia, an area with major resource development projects underway. China, its friendly neighbour to the south, is expanding its infrastructure to access Mongolia's vast mineral resources.

Erdene is a young company, established as a private company in 2002, with an Initial Public Offering completed in March of 2004. Its President and CEO is young as well. Peter Akerley was born in 1963 in Dartmouth, Nova Scotia, just across the harbour from Halifax, earned a Bachelor of Science in geology from Saint Mary's University, and then started traveling the world looking for mineral opportunities. He was building a career in the foreign international exploration business, and what could be more foreign to a young Canadian than British Guyana, Mexico, and the Philippines. He was looking for copper and gold deposits with both major and junior mining companies.

Then, as luck would have it, he found himself in Mongolia in 1997, with "one of the first junior-mining firms to venture into that part of the world." Interestingly, it was fellow Canadians who played a role in the creation and the eventual rise of Erdene in Mongolia: a French-Canadian prospector named Armand Beudoin, who was instrumental in attracting the company Akerley was then with to explore possibilities in Mongolia. The Quebecois invited such firms as International Pursuit and Java Gold, with which Akerley laboured for two years until the Bre-X scandal brought the entire industry to a crashing halt. "My associates and I saw funds drying up, just as we recognized tremendous opportunity in the continued growth of China, along with the mineral opportunities in Mongolia—just as this new democracy was welcoming foreign investment into their nation." Luckily for Akerley and partner Chris Cowan they had formed strong business and personal relationships with two key Mongolian individuals, Bayarsaikhan and Ankhbayar, as well as an Australian based in Ulaanbaatar, Doug McGay. This resulted in the five of them forming a partnership to create Erdene in 2002. Every concept needs its seed money, thus the late Terry Coughlan, a well known east coast mining entrepreneur, and Ken MacDonald, now Erdene's CFO, came into play, assisting with much of the initial financing.

The other Canadian was the world-famous entrepreneur Robert Friedland, who was previously involved in the country, was brought back to Mongolia by his Vice President of Exploration, Douglas Kerwin, who along with Armand Beaudoin, was scouring the country by helicopter looking for opportunities. Friedland's Voisey's Bay discovery in Labrador was the mine which proved that the unromantic mineral called "nickel" can be found in quantities worth over four billion dollars: "the mine find of the half-century," as the promoter has labeled it. The Ivanhoe Group, after evaluating many opportunities in Mongolia, eventually purchased BHP's Oyu Tolgoi prospect in 1999 for $5 million. Today, Ivanhoe's market capitalization exceeds $3 billion, predominantly based on the Oyu Tolgoi deposit hosted in the same belt of rocks as Erdene's Zuun Mod molybdenum project to the west. Things really started happening in the formation of Erdene after Ivanhoe Mines drilled the main discovery hole at Oyu Tolgoi in the summer of 2001. Akerley and his associates created Erdene the following year and began to raise money privately—about $5 million. The fledgling firm then went public on the TSE-Venture exchange at 85 cents, raising another $10 million. Now, they had the wherewithal to be able to begin extensive exploration in the Far East.

After aggressively exploring for a year, Akerley and the Erdene team identified an opportunity to acquire assets in Mongolia including the early stage Zuun Mod molybdenum prospect which were owned by none other than Mohammed al-Fayed. He's the world-class billionaire who owns, among countless other businesses, the world's most famous department store, Harrods, in London. In March, 2005, Akerley and Erdene Vice President Chris Cowan flew across the Atlantic to visit the owner of Harrods and his group in London, and negotiated the crucial deal.

Recalls Peter, "Mr. al-Fayed showed up at five in the afternoon, and I had heard that he was notorious for 11th hour negotiations." But there was no problem: in fact, the billionaire marched in, smiled, said a few words, and quickly closed the deal with his Canadian visitors. Al-Fayed continues to own approximately 3.2% of Erdene.

It was clear from the start that the Zuun Mod project was a very promising one, made all the more enticing by the exploding Chinese demand for steel next door, and the need to strengthen all that steel with molybdenum—and, of course, that unexpected, ten-fold increase in the price of moly during 2005. Akerley suggests, based on current growth, that within four or five years "the annual growth of molybdenum will require the equivalent new production equal to one new major mine every year. That however is where the molybdenum market becomes very interesting, due to the historically low prices during the past decade there was very limited exploration or development. After the price of moly hit $25, existing producers began extensive hi-grading of their molybdenum resources. In other words, they've taken the very best grades of molybdenum from the mines. This process cannot be sustained: it affects mine planning and dilutes the overall grade of the resource moving forward. As a result, we are going to see a decrease in molybdenum supply while demand reaches a critical juncture over the same time frame. I view Zuun Mod as being in the centre of a perfect molybdenum storm: massive increases in Chinese demand and the inability to meet that demand domestically in the next four to five years confronts decreasing supply and grade from the major producers. With three to five years to get Zuun Mod to the production stage, it should be coming on stream at the height of the storm."

Ken MacDonald, Erdene's Vice President, North America, and the company's Chief Financial Officer, is seven years older than Akerley. He is a chartered accountant and a fellow graduate of St. Mary's in Halifax. He adds an intriguing addendum to his friend's words: "Most of this moly is a byproduct of copper and many firms are departing from their mine plan to get the moly out. It's the tail wagging the dog, mining molybdenum because of its high price. They have to get back on track to get copper extraction on the agenda." Notes Akerley, "Two of the top five moly producers have announced that they're going to have to scale back their production of molybdenum because of this depletion of their mines by taking away the high grade pockets of moly for the profit. Now, they'll be forced to mine lower grades in the future and will have to face the music."

What Akerley and others quickly discovered in Mongolia—and was obviously evident to a billionaire businessman in London, England—was that Zuun Mod was a solid flagship property that was worth acquiring. After drilling only a few holes, they found molybdenum throughout a 4.5 kilometre by 5 kilometre area. "It is quite substantial for any resource to be spread over that extensive an area," enthuses Akerley. This led their geologists to drill more, and they soon "intersected molybdenum from near-surface down to the bottom of the deepest hole we've drilled—430 metres—finding grades from 0.05% to 0.07% molybdenum over widths greater than 100 metres. And some of our most recent drilling shows higher-grade intersections within those zones." What it all indicates to geologist Akerley and his colleagues is "the potential for a resource that could be world class."

In mid-June, 2007, Erdene announced in a press release stating it has completed the planned Phase 1 drilling program at its wholly owned Zuun Mod molybdenum project in Mongolia. Results received from drill holes ZMD-28 and ZMD-29 confirm that the molybdenum mineralization previously demonstrated to be 380 metres thick (ZMD-13, 0.05% Mo) in the Racetrack Zone continues over a minimum strike length of 1.4 kilometres and comes to within 18 metres of surface. This effectively triples the previous known extent of this zone and suggests possible continuity along the entire 3.2 kilometre-long South Corridor of the Zuun Mod porphyry complex.

"The significant extension of strike length, locally highmolybdenum grades, wide intersections and molybdenum mineralization coming much closer to the surface are very exciting developments for the Zuun Mod project, greatly enhancing the economic and resource potential," said Peter Akerley, President and CEO. "With the pending development of major metal and energy projects across Mongolia's Gobi region, Zuun Mod is well located to benefit from the associated infrastructure and service build-up, as well as its proximity to China's processing centres and molybdenum markets."

These promising results have moved Akerley and his management to put two rigs back on site to continue the drilling. "We don't know how big it is yet, as we continue to expand its limits," he explains. "We believe, based on this drilling, that we have the potential to far exceed our goal of defining 250 million tons of 0.05% moly." In fact, in an August 9, 2007 news release, Erdene reported additional results indicating the area of known mineralization continues to grow. In the central part of the deposit new holes have extended the east and west margins of the mineralization with 0.07% to 0.08% molybdenum over sections averaging greater then 100 metres in thickness and within mineralization remaining open at depth.

"Once we complete this next round of drilling ," Akerley declared, "we anticipate to be in a position to release a resource estimate for the Zuun Mod property. We feel Zuun Mod has the potential to host upwards of 300 million pounds of molybdenum. Zuun Mod's mineralized zones coming to within 18 metres of surface is a very positive factor in the overall economics." Akerley, along with many other mining executives and experts interviewed for this book, is confident that prices for that more-and-more-precious metal will remain in the $25-plus range over the long term. It's good to report that Erdene's own benchmark for their predictions and expenses has been based on price estimates of $15-a-pound molybdenum.

While Erdene has additional mineral interests in both Canada and the U.S., "the big-dream discovery," to use the phrase of the firm's President, remains in the Mongolia region. It's really about the China minerals boom, and Akerley sees many billions of dollars in investments pouring into that great and populous land over the next several years. "We are in the perfect place to be developing a mining project next door to China. This under explored geologic terrain now hosts two massive porphyry molybdenum, copper and gold deposits newly discovered in the past 10 years," Peter declares. "On a comparable basis," Akerley adds, "when you look at the other companies with moly projects in the development stage and you do not take into account the company's other projects, we could very well see an increase of our share price, based on the value of the Zuun Mod project alone. Certainly, we would expect a positive increase in the share price within the next year, and even more so in the years that follow, as we firm up the potential of Zuun Mod." The President admits that there are a few good projects on the world's molybdenum drawing board, the Zuun Mod project being only one of them, but "you can't just pick these off the tree," he insists. "There are not dozens of world-class moly deposits out there and those of substance will continue to become fewer as we rapidly deplete the ones we have. Zuun Mod is unique in its location relative to the growing market; it also is the eventual beneficiary of a rapidly developing mining industry in the Southern Mongolia region."

What they have also discovered is a by-product called Rhenium, a metal first identified relatively recently in 1925 by German chemists; hence the name, taken from the Latin word for the River Rhine. Currently selling for $2500/lb, rhenium has been one of the best performing precious metals over the past few years. The reasons for this include the demands placed on the metal by the the U.S. military, which needs rhenium to produce parts for the control of the exhaust of the very high temperature jet engines used in stealth aircraft. In addition it has been discovered that rhenium diboride can be made easily by simply reacting elemental rhenium and boron and that the resultant compound is harder than diamond which has increased markets for the metal in specialty tools and complex metals. Erdene is aware of only a handful of mines which contain enough Rhenium to make it worthwhile to recover. Zuun Mod, based on early analysis, has significant amounts of Rhenium, so Akerley and his fellow managers plan to continue to investigate its potential value. At $2500/lb, it's "quite significant."

Mines, like Rome, are not built in a day. While some of Erdene's projects in the United States are already in production, the Zuun Mod project in Mongolia is still in the advanced exploration stage. It may take three to five years to reach initial production, assuming all continues to go well in pre-feasibility and feasibility studies. In our world of 1.3-second downloads and instant gratification, several years can seem like an eternity, but the potential value of the Zuun Mod project is obviously enormous.

"Based on what we set out to do with this most recent drill program," explains Akerley, "we were looking for a certain area that would provide us with the tonnage necessary to justify a large, 12 million pound-plus per annum operation. We believe that we've achieved that goal and have gone beyond, and extended the previously known limits of the resource area. However, we must conduct additional drilling to determine the extent of the molybdenum mineralization at Zuun Mod. That's why we're quite excited about the tonnage potential indicated by our most recent round of drilling. I've always looked at a 20-plus year lifespan for a quality opportunity, and in the Gobi Desert, we'd like to see over 30 years. The major mining companies look for world-class, 30- to 40-year operations."

A major shareholder of Erdene—owning more than Mohammed al-Fayed—is Xstrata, at 4.5%. The latter is a major global diversified mining group listed on the London and Swiss stock exchanges. Akerley is dealing with their coal office in Sydney, Australia. This relationship is more focused on the coal side of Erdene's business in Mongolia, and the firms are involved in several projects together. But Akerley and his directors have no plans to look at partners for Zuun Mod until they firm up Erdene's value for their shareholders.

The potential for joint ventures is always out there, and "some people and firms are certainly interested," according to Akerley, in both partnering with Erdene and/or purchasing parts of the firm. But he and his team believe they have the financial and human resources necessary to take the project to the next level, and they have every intention of moving forward with 100% ownership of Zuun Mod.

Although a visitor in Mongolia, Erdene believes the development of mineral projects can have significant benefits to not only the shareholders of Erdene but to the people of Mongolia. The entire population of the geographically large nation is only 2.7 million, less than those living in the Greater Toronto Area. Its capital of Ulaanbaator has fewer than 750,000 souls, about twice the number of the citizens of Halifax, Nova Scotia, where Erdene has its head office. Nearly one-third of the population of Mongolia is nomadic, something that Western mining companies must be sensitive to.

"We must do everything we can to ensure that we cause the least amount of interference with this very unique culture," insists Peter Akerley. It is reassuring to read on Erdene's website that, because of Mongolia's "well-preserved and unique ecology," the firm's objective is "to minimize our environmental footprint," filing environmental protection and reclamation plans with the governor of each district in which the firm operates, working to ensure those plans exceed requirements. The company places a priority on hiring local workers whenever possible, and prides itself on its upgrading of facilities at an orphanage in the capital city, funding scholarships, providing financial support for students who are undertaking graduate work at Ulaanbaator's universities and providing hospital and school supplies to villages located near their exploration properties.

As of the summer of 2007, Erdene was asked whether they were progressing at the speed and in the direction that Akerley and his directors had hoped. "Absolutely!" Peter Akerley responds. "We'll spend somewhere in the vicinity of $8 million on exploration and development projects and the bulk of that investment in Zuun Mod. Recent results are better than we had expected, and we truly feel that we have a world-class opportunity on our hands." Erdene's management must be pleased that they have $14 million in cash (summer 2007), which means they don't have to offer more shares and dilute the present shares' value. There are fewer than 66 million shares outstanding, a number with which Akerley is pleased. The firm raised $10 million only recently, at a dollar a share. The ability to go back and raise additional funds if necessary is certainly there, but the Erdene board would prefer to do that when further value is recognized in their share price through Zuun Mod.

In addition to the Zuun Mod project, the company is creating value through exploration of a number of other projects in Mongolia. They include the Biger Copper-Gold-Platinum Project; the Mogoit and Tsagaan Ovoo Copper-Gold Projects, the Erdenet Copper Project, the Galshar Coal Project (along with Xstrata Coal Canada Limited), all of them sprinkled throughout Mongolia. Then, there is the Donkin Coal Project in Nova Scotia, Canada (again partnered with Xstrata Coal in a 25:75 joint venture) and the Maddox Aggregate and the Sparta Kaolin Projects in Georgia, U. S.A.

Erdene has a number of impressive people in senior management, and both Akerley and MacDonald eagerly mention Chris Cowan, who spent three decades of his career with Falconbridge, the then-Canadian mining giant (which was later purchased by Swiss mining giant Xstrata PLC in the summer of 2006), managing several of their large operations. He is now running Erdene's Mongolian operation and is considered "a key individual in all aspects of our business." Erdene's President and its CFO also feel they have an excellent team of geologists in place, with a small staff of seven at the head office in Halifax, along with 13 Mongolians and three ex-pats employed in their Ulaanbaatar office. Including contractors, the number of people in the Far East is five to six dozen during the field season, hitting a peak in the late summer of each year.

Asking a President and CEO about his corporate philosophy usually gets the standard reply of "to create value for our shareholders," but Peter Akerley goes a lot further in his reply: "The best way to improve our stock price is to identify unique opportunities in the mineral business, and to use the relationships we have built up over the past two to four decades to find the right deals. Between myself, Ken MacDonald and Chris Cowan, we have 80 years of experience in the business, and none of us is quite ready to retire. We believe that we have the ability to bring more projects on board that others may not have identified, and to create great value in those, before we look for any partnerships." One usually thinks that success in this business is spelled "m-i-n-e-s", but not according to Peter Akerley. "It's p-e-o-p-l-e"; no question, it's people who make a successful company. Good geologists find deposits but a multi-disciplined team with strong chemistry is required to support that field geologist and to ultimately take that deposit and create value."

The President of Erdene Gold smiles when he notes that he and his team have taken an exploration company worth $5 million less than four years ago, to the $80 million-plus it is worth today. "And we expect terrific upside. That is very rewarding," he nods. In terms of "Moly in Mongolia," he is not aware of any other project that rivals Zuun Mod, or even comes close to it. Worldwide, there are a few other firms out there which could be considered competition to Erdene. "When you look at the new annual demand for moly hitting over 20 million tons by 2012 - 2015, there's lots of room at the table, and the competition shouldn't be too fierce," Akerley predicts.

In creating value for Erdene's shareholders, Akerley looks to the East, because it is with Zuun Mod of Mongolia where they have the greatest hope since "it has the potential of being a world-class molybdenum mine with an excellent location relative to market and a new mining industry growing in the region. The truth will be at the end of the drill, as we wind up the present drilling and move forward from there." He has no doubt that his young company, Erdene (meaning "treasure" in Mongolian), is sitting on a resource that rivals some of the largest, with a stock price sure to perform well over the coming years, as current and future shareholders come to recognize the unrealized value of Zuun Mod and Erdene.

CORPORATE SNAPSHOTS

ADANAC MOLY CORP.

Stock symbol: AUA – TSX

$1million new bridge across Pine Creek accessing the Ruby Creek project

Core samples from recent drill program

Placer Developments ore samples from 1980

Ruby Creek overhead view

Head Office: 2A – 15782 Marine Drive
White Rock, BC V4B 1E6

Telephone: 1.604.531.9639

Website: adanacmoly.com

BARD VENTURES LTD.

Stock symbol: CBS – TSX

Close proximity to Hydro line in Houston, BC

Company personnel review drill core

Drill core Shack

Lone Pine drill core

Head Office: 1255 W. Pender Street
Vancouver, BC V6E 2V1

Telephone: 1.604.687.2038
Website: bardventures.com

COLUMBIA YUKON EXPLORATIONS

Stock symbol: CYU – TSX

Douglas Mason, Chairman of Columbia Yukon Explorations Inc., and Kevin Carlick, of the Dease River First Nation, shaking hands

Jill Pardoe, government inspector, inspects drill at Storie

Storie Deposie showing drill hole locations

Storie Mountain

Head Office:	2489 Bellevue Avenue West Vancouver, BC V7V 1E1
Telephone:	1.604.922.2030
Website:	columbiayukon.com

COPPER FOX METALS INC.

Stock symbol: CUU – TSX

Copper Fox\HiTech's SuperDrill in action

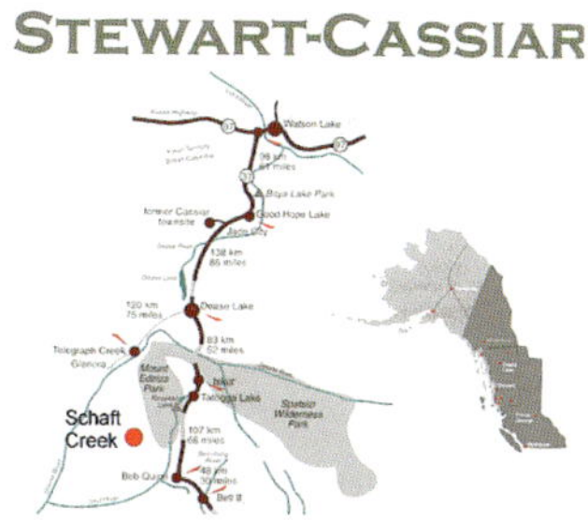

Map of Schaft Creek property

Old Claim Tag (circa 1957) for SNO1 Claim

Walter Hanych, Geologist, examing core samples

Head Office: 650, 340 – 12th Ave SW
Calgary, AB T2R 1L5

Telephone: 1.403.264.2820
Website: copperfoxmetals.com

EAGLE PLAINS RESOURCES LTD.

Stock symbol: EPL – TSX

Exploration camp, Fall 2006

High-grade moly mineralization in Sphinx drill hole

Sphinx Drill Pad, May 2007

Sphinx Property Area - Drill pad in centre of clear cut

Head Office: Suite 200, 16 – 11th Ave S.
Cranbrook, BC V1C 2P1

Telephone: 1.250.426.0749
Website: eagleplains.ca

ERDENE GOLD INC.

Stock symbol: ERD – TSX

Zuun Mod is strategically located on the doorstep of China's booming steel industry

Molybdenite in drill core from Zuun Mod project - Phase I resource delineation drilling, July 2007

Mongolian boy preparing for horse racing at the traditional Naadam Festival

Peter Akerley (far left) and the Erdene team discuss exploration results

Head Office: 99 Wyse Road, Suite 1480
Dartmouth, NS B3A 4S5

Telephone: 1.902.423.6419

Website: erdene.com

FORTRESS BASE METALS CORP.

Stock symbol: FBM – TSX

Davidson project on Hudson Bay Mountain property

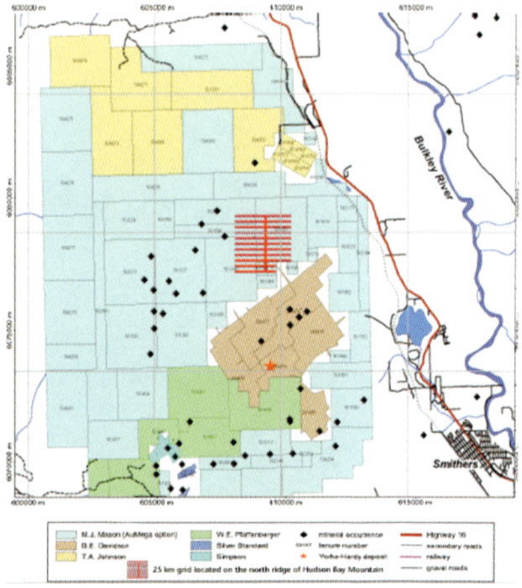

Drill on Poplar Lake

Poplar property location map

View to the southeast over Poplar Lake

Head Office: 2nd Floor, 157 Chadwick Court
North Vancouver, BC V7M 3K2

Telephone: 1.604.988.5336

Website: aumegadiscoveries.com

GEODEX MINERALS LTD.

Stock symbol: GXM – TSX

The Geodex crew processes drill core at the Nackawic core facility

NB Field Meeting, August 2006

The Geodex group discusses progress on the Sisson Brook property

The 2007 trench on Zone III at the Sisson Brook property

Head Office: 800 W. Pender St., Suite 450
Vancouver, BC V6C 2V6

Telephone: 1.604.689.7771

Website: geodexminerals.org

GOLDEN PHOENIX MINERALS INC.

Stock symbol:
GPXM – OTC:BB

Ashdown Ball Mill

Ashdown crusher

Ashdown portal

Miners with a serious moly sample

Head Office: 1675 E. Prater Way, Suite 102
Sparks, NV 89434
Telephone: 1.775.853.4919
Website: golden-phoenix.com

HAPPY CREEK MINERALS LTD.

Stock symbol: HPY – TSX

Discovery Moly Zone 3

Discovery Moly Zone

Looking west from Deception Mt to Deception Lake

Ridley Creek zone- new outcropping moly zone

Head Office: 1066 W. Hastings St., Suite 2304
Vancouver, BC V6E 3X2

Telephone: 1.604.662.8310
Website: happycreekminerals.com

HI HO SILVER RESOURCES INC.

Stock symbol: HIHO – CNQ

Fred Fisher, President of Hi Ho Silver

Fred Fisher and Isabel Alves

South Rim gold moly project

Tasco copper moly project

Head Office: 3045 Southcreek Rd., Unit 11
Mississauga, ON L4X 2E9

Telephone: 1.905.602.4653

Website: hhsr.ca

INTERNATIONAL PBX VENTURES LTD.

Stock symbol: PBX – TSX

Claudio Burgoa & Miquel Guerrero_PBX Geologists

Cerro Moly Molybdenum Mountain

Drilling top of Cerro Moly Molydenum Mt

Horizontal drilling into side of Cerro Mt.

Head Office: 475 Howe St., Suite 209
Vancouver, BC V6C 2B3

Telephone: 1.604.681.7748

Website: internationalpbx.com

KOBEX RESOURCES LTD.

Stock symbol: KBX – TSX

Rom Shklanka, Chairman of Kobex Resources

Leo King, President of Kobex Resources

Lucky Jack property

Geologists looking at the core

Head Office:	700 W. Pender St., Suite 1700 Vancouver, BC V6C 1G8
Telephone:	1.604.484.6228
Website:	kobexresources.com

MOLY MINES LTD.

Stock symbol: MOL – TSX / MOL – ASX

Brendan Cummins and Mark Allen inspecting core

Drill core with moly evident

Miss Molly the Moly Mines mascot

View to the west over resource area

Canadian Office: 80 Richmond St. W., Suite 1801
Toronto, ON M5H 2A4

Telephone: 1.416.777.1801

Website: molymines.com

MOSQUITO CONSOLIDATED GOLD MINES

Stock symbol: MSQ – TSX

High Grade Molybdenite at Cumo

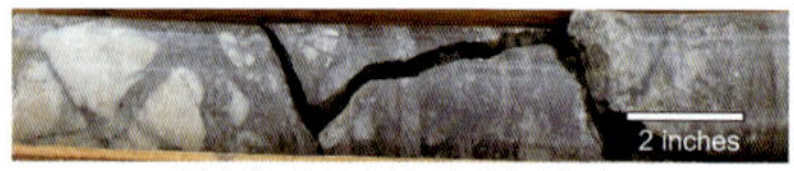

High Grade Molybdenite (Pine Tree)

Molybdenite-quartz (Pine Tree)

Molybdenum Stockwork (Pine Tree)

Pine Tree core samples

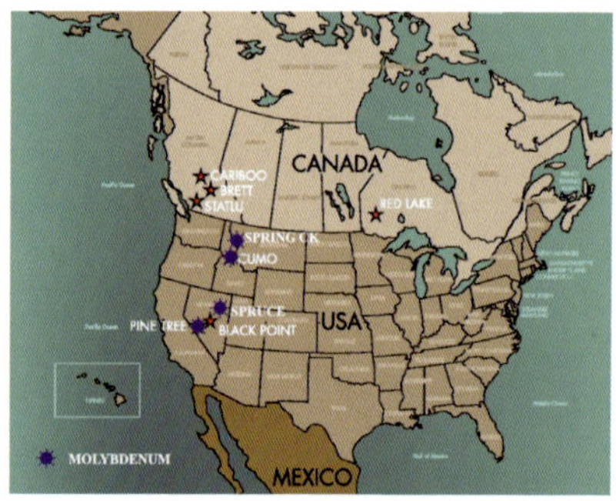

Map showing molybdenum properties

Pinetree drill

Head Office: 455 Granville St., Suite 301
Vancouver, BC V6C 1T1

Telephone: 1.604.689.7902

Website: mosquitogold.com

PACIFIC BOOKER MINERALS INC.

Stock symbol: BKM – TSX

Bell Copper mine

Locations of former producers

Overview of Granisle Village

Pacific Booker Minerals camp, 2000

Head Office: 1166 Alberni St., Suite 1702
Vancouver, BC V6E 3Z3

Telephone: 1.604.681.8556
Website: pacificbooker.com

QUADRA MINING LTD.

Stock symbol: QUA – TSX

Malmbjerg 200 Camp at Sunset

Morris Beattie at Malmbjerg camp, June 2007

View from Malmbjerg looking past plantsite towards Port Site

View of Malmbjerg deposit from the west

Head Office:	1177 W. Hastings St., Suite 2000 Vancouver, BC V6E 2K3
Telephone:	1.604.689.8550
Website:	quadramining.com

ROCA MINES INC.

Stock symbol: ROK – TSX

MAX Aerial site, October, 2007

Messrs. Skerlec, Broughton, and Mirko recieve CMEBC Award, November, 2007

MAX Flotation cells, October, 2007

MAX MoS2 flotation concentrate, October, 2007

Head Office:	1122 Mainland St., Suite 490 Vancouver, BC V6B 5L1
Telephone:	1.604.684.2900
Website:	rocamines.com

SULTAN MINERALS INC.

Stock symbol: SUL – TSX

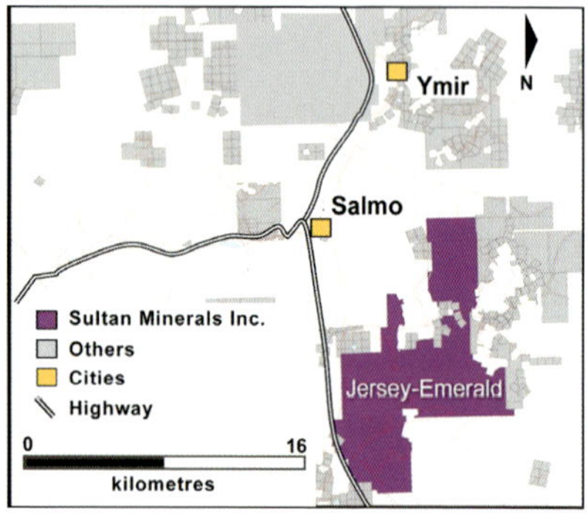

Area map of the Jersey Emerald property located just south of Nelson, B.C.

Art Troup, President and CEO of Sultan Minerals

Core samples showing high grade molybdenum

Ed Lawrence showing the molybdenum intersected in one of the underground tunnels

Head Office: 570 Granville St., Suite 1400
Vancouver, BC V6C 3P1

Telephone: 1.604.687.4622

Website: sultanminerals.com

TERRANE METALS CORP.

Stock symbol: TRX – TSX

Aerial photograph of the Mt. Milligan exploration camp site

Mt. Milligan drill hole core, 2007

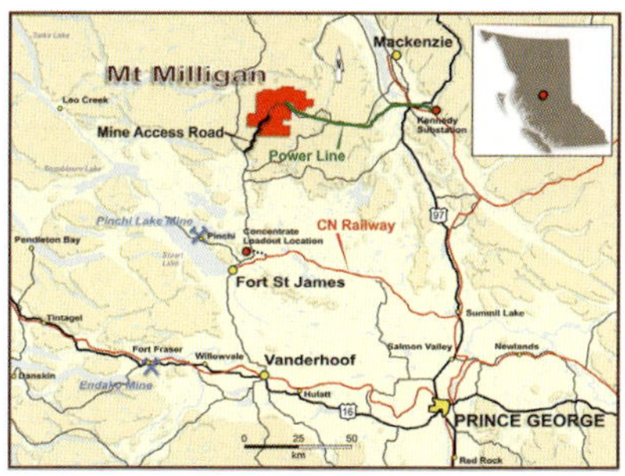

Mt. Milligan location map

Terrane President and CEO Rob Pease presenting details on Mt. Milligan site layout

Head Office:	999 W. Hastings St., Suite 1500 Vancouver, BC V6C 2W2
Telephone:	1.604.681.9930
Website:	terranemetals.com

TORCH RIVER RESOURCES

Stock symbol: TCR – TSX

CEO of Torch River, Bill Pfaffenberger

Don Davidson at Red Bird

Red Bird drill site, elevation 1445 metres, July, 2007

Red Bird property

Head Office: 888 – 3rd St. SW, Suite 1000
Calgary, AB T2P 5C5

Telephone: 1.403.444.6888
Website: torchriver.ca

WESTERN TROY CAPITAL RESOURCES LTD.

Stock symbol: WRY – TSX

High-grade molybdenite in a South Zone boulder

MacLeod Lake Property map

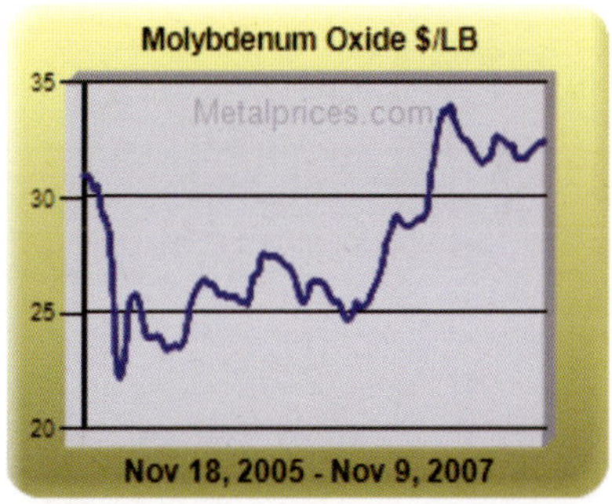

Moly Two-year Price Chart

Molybdenite from the Northeast Zone

Head Office: 102 Bloor St. West
Toronto, ON M5S 1M8

Telephone: 1.416.929.3268

Website: westerntroy.com

WIN-ELDRICH MINES LTD.

Stock symbol: WEX – TSX

Ashdown Gold Mine over a century ago

High grade moly ore from the Sylvia Vein

Massive molybdenum exposed at face of 1983 sill drift

Moly stockpiled in 1983, milled, concentrated and sold by Win-Eldrich in 2005

Head Office: 4 King Street West
Toronto, ON M5H 1B6
Telephone: 1.416.363.1954
Website: win-eldrich.com

FORTRESS BASE METALS CORP.

Chadwick Wasilenkoff was "the man to see" when it came to restructuring a Canadian-based junior resource company by the name of 'Aumega Discoveries Ltd.' The resulting Fortress Base Metals Corporation is a reincarnation of the previously gold-centric company, which now has no qualms about leaving its precious metal roots behind. "I would say we're definitely a base metal company: copper and molybdenum. There happens to be some gold on our properties, but it's the copper and moly we're after," says Wasilenkoff, the Vancouver company's de facto head.

Molybdenum is certainly one of the most important metals in Fortress's inventory. Focused on the exploration and development of both advanced and early stage mineral projects throughout British Columbia, Fortress currently has a slate of promising deposits, including some with very interesting moly values.

Fortress Base Metals expects big things from its Poplar property, in which it has the option to earn a 100% interest. This advanced stage copper–molybdenum porphyry deposit is located 75 kilometres south of Houston, BC. Fortress also already holds a 100% interest in two other B.C. properties—Kelly Creek and Hudson Bay Mountain. Kelly Creek is a high-grade copper–silver prospect located 28 kilometres southeast of Terrace; and the Hudson Bay Mountain property is a collection of mineral tenures 13 kilometres northwest of Smithers. The tenures that make up Hudson Bay Mountain surround Blue Pearl Mining's Davidson molybdenum deposit (formerly known as the Yorke-Hardy deposit), which is hosted in the Glacier Gulch intrusive complex. "And there are numerous occurrences on our property," confirms Wasilenkoff. "There's been some preliminary work done that gives a strong indication that these porphyry moly deposits often occur in pods. And Blue Pearl's deposit has multiple high-grade pods. We feel those pods probably continue on to our property, both to the north and to the south."

With a trio of encouraging properties, Fortress has a fresh, positive outlook, but the company's objectives remain true to those of Aumega Discoveries when it first acquired the Poplar and Kelly Creek, back in 2004. "I think for the company, the underlying properties have been maintained and the objective there stays the same. We've got a great portfolio of properties," states Wasilenkoff.

"Our most advanced, or main deposit is our Poplar deposit. We feel it probably needs one to two more rounds of drilling—basically, more confirmation drilling. The Poplar is a well-known deposit, and we feel it could be ready, and could fairly quickly go into production. In terms of production, typically—especially in the British Columbia environment—it can take five to six years to get there. We feel the Poplar, potentially, could get into production in less than two years."

One reason Fortress Base Metal's expects to get Poplar into production so quickly, Wasilenkoff notes, is that "there's an existing mine only 35 kilometres away. The Huckleberry copper-molybdenum mine has less than one year's worth of ore and reserves left, but they have excess capacity in their tailings pond, and they've got all their permits in place." Wasilenkoff explains that one of the most challenging aspects of mine development is building and permitting a facility like the one at Huckleberry. It's much easier to get permits for exploration and transportation. Which is the stage Fortress is at with the Poplar deposit. The rare opportunity to transition an existing mine operation, rather than construct a whole new mine site, is like the icing on the cake at Poplar. Poplar lies within the strongly mineralized Intermontane Belt, east of the coast Crystalline belt. These rocks are cut by numerous porphyritic intrusives of Jurassic, Cretaceous, and early Tertiary age, which host several significant porphyry copper–gold–molybdenum deposits, such as Huckleberry, Whiting Creek, Berg, Ox Lake, and others. There are several stocks of differing composition outcropping within the sixty-two claim blocks that make up the Poplar property. Early drilling at the site was only carried out in the Poplar stock, in the Canyon Creek area, although the other stocks, such as the China Creek stock, also have potential for porphyry mineralization.

The Poplar property was originally staked in 1971 by Frank Onucki and partners for El Paso Mining and Milling Company, who carried out soil geochemical surveys, geological mapping, and bulldozer trenching in 1971 and 1972. Yet, El Paso abandoned the option at the end of that year, before finding the porphyry-style mineralization. In 1974, the property was optioned from the vendors by Utah Mines Ltd. Work carried out by Utah included geological mapping, soil geochemical surveys, and magnetometer and IP geophysical surveys.

Up to 1982, Utah completed a total of 17,900 metres of diamond drilling in seventy-three drill holes. A full-scale positive feasibility study was completed by Utah in 1980, followed by other studies until the end of 1982. But as metal prices were sliding, Utah faced a major final payment and they ultimately chose to hang on to their money and return the property to the vendors.

"Utah did a feasibility study on the copper side," Wasilenkoff says, "and the only thing that forced them to walk away from the property at that time was the poor copper price. Back then copper was less than a dollar a pound. We feel that the Poplar could probably be economic in the $1.25 to $1.50 range. Even at those prices it could still be a profitable opportunity. Obviously with a $3 copper price, it becomes very interesting and exciting. It's a low- to medium-grade deposit, but at $3 plus per pound of copper, there's quite an upside."

In 1991, New Canamin Resources Ltd. acquired the Poplar property and completed an additional 13 short diamond drill holes totaling 1300 metres. Total expenditures on the property by Utah, Canamin, and others, exceeded $2 million. As a result of all that work, various resource, tonnage, and grade estimates have been made by major exploration companies. The most commonly quoted estimate is 116 million tonnes grading 0.32% copper, 0.1 g/t gold, and 0.009% Mo (at a 0.20% copper cut-off grade). Appreciable rhenium values are also contained in the molybdenite. This resource, although prepared prior to the implementation of NI 43-101 standards, is considered a reliable estimate comparable to an indicated mineral resource in accordance with Canadian Institute of Mining definitions.

Within the 116 million-tonne resource, the additional 1991 drilling permitted the estimation of a smaller but higher-grade near-surface mineral resource of about 40 million tonnes of 0.40% copper, which could likely be expanded even further with extra drilling. "Poplar would definitely be a copper project, with gold and molybdenum credits," says Wasilenkoff.

Potential also exists for several hundred million tonnes of mineralization in Poplar's combined Main and East zones. Many of the deep drill holes encounter the best grades at depth and terminate in mineralization with good copper grades. The deepest hole, DDH-PC-35, terminates in mineralization at 609 metres.

Consulting geologists involved in the 1991 drill program also felt that there was plenty of potential in the China Creek stock for a new porphyry Cu–Mo zone. Mineralization comparable with that at the Huckleberry East zone could be hosted in the andesite wall-rock of the China Creek porphyry body. The three drill-holes completed to date on this zone are not considered to adequately test the area, and have not been able to explain the copper soil anomalies in the area. Therefore, China Creek remains a target to be tested in future exploration, particularly by IP geophysical surveys.

"Now we're talking with some of our geologists that have written up reports, and done the evaluations," says Wasilenkoff. "They feel that the Poplar deposit is quite a bit larger than what we're able to include based on current data in terms of an NI 43-101, or indicated resource. The geologists believe the deposit is over 400 million tonnes because it's still open at depth, and in several directions.

"Just to pull up some of our drill holes, for instance: we hit 42.9 metres of 0.62% copper and 0.15 g/t gold; 166 metres of 0.41% copper, 0.12 g/t gold and 0.12% Mo. So, again, there are definitely some higher-grade intervals there. We've also done some three-dimensional modeling to get a better understanding of the opportunity. And as I said, our general geological plan is to probably have one more round of drilling, potentially two more rounds of drilling by the end of 2007. It'll be in the neighbourhood of four to six holes each. Probably less than a million dollars, and then we're going through the feasibility study quite quickly."

But the Poplar is only one of Fortress's three properties. "We also have two very exciting properties, which are more in the exploration stage, but still, properties with exciting potential: our Hudson Bay Mountain and Kelly Creek," says Wasilenkoff.

Whereas, in leaner years, the molybdenum at Poplar might have been passed over, it's a very different story at Fortress's Hudson Bay Mountain property, where the moly producing potential is undisputable. "Absolutely, the focus will be molybdenum and trying to find an extension of the Davidson deposit." The Hudson Bay Mountain property is comprised of 46 mineral tenures, containing 659 cell units, and covering 12,278 hectares. The deposit itself is reported to have a measured resource of 6,070,000 tonnes grading 0.286% MoS2, and an indicated resource of 102,610,000 tonnes grading 0.259% MoS2, both at a cut-off grade of 0.17% MoS2.

The main target for future exploration programs will be uncovering additional molybdenum resources related to the Glacier Gulch intrusive complex. "Basically, on the Hudson Bay property, we want to complete a bit more ground work and do some airborne to establish some drill targets. We do have areas of some past production on the property. They are small pockets of mineralization in veins, and we feel those veins are probably associated with deeper lying and underlying porphyry systems, similar to the Davidson deposit. With some aeromagnetics, we'll be able to better define those and set up the drill program. I'm not sure if we'll be able to drill that prior to the end of 2007; if not, it'll happen in early 2008," says Wasilenkoff.

Meanwhile, the Kelly Creek copper-silver project is also situated in B.C., in an area covered by mineral claims Kelly 1 through Kelly 4 and K.C.1 through K.C.4. The claims are comprised of eighty-four mineral claim units within one contiguous block. The total property area is 2,000 hectares. The property is accessible from Terrace, 30 kilometres west via British Columbia Highway 16 and a main logging road. Four-wheel drive roads provide access to the southern and central parts of the property. Fuel, supplies, and scheduled air, ground, and rail transportation are available at Terrace.

Previous exploration within the area now covered by the Kelly Creek project is extensive, though not all the results of past work are available. Work includes prospecting, trenching, soil geochemical surveying, IP geophysical surveying, 302 metres of underground workings, and both underground and surface diamond drilling. According to a 2005 property report, drill holes in the Upper Kelly Creek prospect area contained up to 5.35% copper and 95.3 g/t silver over 4.6 metres (drill hole UK5), and 1.30% copper and 28.8 g/t silver over 34.5 (drill hole UK10).

Historically, in 1981, Cathedral Minerals Ltd. estimated a resource at the Upper Kelly Creek prospect consisting of an indicated 2.27 million tonnes containing 1.03% copper and 18.5 g/t silver, and an inferred resource of similar tonnage and grade. This estimate was performed prior to the introduction of current resource reporting standards. "There's been some past production; there's been significant drilling on the property, with some very high grades throughout the veins. So, we'll do some airborne surveys, just to better define our drill program, but we'd like to drill Kelly Creek as well by the end of the year or early next year."

The Kelly Creek property is within the Intermontane Belt of the Canadian Cordillera near the western contact with the Coast Range Plutonic belt. The property is also prospective for volcanic-hosted "red bed" copper deposits. Red bed usually refers to layers of reddish sedimentary rocks, which were deposited in hot climates under oxidizing conditions. But in this case, the red bed is associated with the volcanics of the Hazelton Group, and is indicative of copper porphyry deposits and precious metal deposits associated with granitic intrusives. The Upper Kelly Creek deposit also has many similarities to volcanic red bed copper deposits—deposits such as the Calumet Mine in Michigan, where 72.4 million tonnes were produced containing 2.64% copper, and the Sustut deposit in B.C., where a resource of 8.24 million tonnes containing 1.62% copper has been identified.

"As you can see, all three properties are at different stages, which is great," says Fortress Base Metals' interim President Chad Wasilenkoff. "So, as one of these comes to fruition, the plan is that we will continue to add more properties in the future. But, for at least the next twelve months, we've got enough on our plate with these three."

With BC's low political risk, and minimal environmental and First Nations complications, the overall dynamics seem to be working out as well. "We have none of those issues," says Wasilenkoff with a wave of his hand. "And the general logistics—all of these Fortress properties are within about 100 kilometres of each other. So, when we send people up there, they can work on all three different properties, and utilize the entire team."

Now at a stage where everything is advancing, without leaving any of their projects waiting in the wings, Fortress Base Metals represents the best of both worlds to their shareholders. "Certain investors like just 'blue sky' and exploration. Others want to see something proven before they would get involved. We offer both." Wasilenkoff continues, "We have a proven resource at the Poplar that we feel can get into production fairly fast; so there's limited drilling risk. There's always risk when trying to bring these things into production, and then underlying commodity price risk. But, we also have these very large opportunities, or potential 'blue sky' opportunities, with the Kelly Creek and our Hudson Bay Mountain deposit."

Joint ventures are an excellent tool for mitigating the risk of developing a property. And Fortress is definitely open to working with other companies. "We acquired our Hudson Bay Mountain property very shortly after Blue Pearl acquired the Davidson," Wasilenkoff says. I would say, basically, we were on par with those guys in trying to build a molybdenum company." Which places Fortress in a prominent position to joint venture should the opportunity arise; "Yes, we would look at joint venturing with anybody, as long as they have a strong reputation and some expertise to bring to the project. The Blue Pearl guys have a very strong knowledge base and a good understanding of the exploration model they've been looking at up there. So, we would definitely look at partnering with them on that particular property.

"There are other groups that are more experienced on the copper porphyry types of deposits we have, like the Poplar," Wasilenkoff says, "therefore we would look to partner with them on that property. Imperial owns the

Huckleberry mine. But we've had no discussions with any of these groups to date. We're always open to a creative business relationship."

For Wasilenkoff, and for Fortress's investors, the hard work of restructuring, renaming, and changing the focus of Aumega is starting to pay off. "We basically went through an entire restructuring," explains Wasilenkoff. "We changed the entire board of directors, management; everything right across the board. We rolled the stock back, five for one. We just wanted to get the company back in good standing, to rebuild the structure and basically have a good platform to move forward from."

These kinds of corporate evolutions are Chad Wasilenkoff's specialty. "My role in these types of things, in the mining world typically, is to set the company up, get it organized, put the properties together, build a management team, and then assist after that, to stay on as an active board member," he says. "So, the majority of my role at Fortress is now done. All the pieces are put together."

Wasilenkoff first became involved when he joined the board as a Director from July 2004 until June 2006. "I was President and CEO of Titan Uranium Inc.," he notes, a company he also helped restructure. And I've got another gold company in Ecuador that I've helped set up. I remain on the board of advisors, but that was set up about five or six years ago." But, "Titan and Fortress; these companies are very, very different. My role in Fortress Base Metals, back when it was Aumega Discoveries, was as a director. It was supposed to represent less than 5% of my time. And again, just in a director role, I was trying to help with corporate development and strategy and things like that. Then the company got themselves into trouble; they'd overspent. They had some problems with maintaining their ground—losing it to stakers. They were in debt, management was not very focused, they were having a difficult time raising money. But, at the same time, when you looked at the portfolio of properties, it was clearly a spectacular company. And the market capitalization of the company, at the time, was ridiculously low. So, that's when I made the decision to step in, clean house, and rebuild. "This included the name change, the roll back or consolidation of shares, and basically putting the company in a position to make it attractive as a public company, because we know it's attractive as a mining company."

Now in his mid-thirties, Chadwick Wasilenkoff already has a considerable breadth of business experience. He has been President and CEO, and a Director, of Fortress Paper Ltd. which makes security paper for the Swiss Franc, the Euro, and passports for numerous countries. Fortress Paper recently completed a $46 million IPO on the TSX which was led by RBC and TD Securities. Mr. Wasilenkoff also keeps his hand in a number of other private business interests, and holds a degree from the University of British Columbia he completed in 1995.

"I've always been interested in business in general. And I've always been a contrarian type of investor," he says. "For example, we built a gold company back when everyone else was doing tech deals. We bought the asset for a million dollars; the company's now worth about a quarter billion dollars. Just migrated it to the TSX big board, trading at about $9, we started the company at 15 cents! And then, we went into uranium back when nobody was looking to invest in uranium. We got that asset out of Cominco for $10,000 and built a uranium company – Titan.

"Now we're doing the same thing. When we picked up these copper and moly assets, nobody was really following the copper or moly markets. The increase was just getting going. Because we're just re-launching the company right now, it looks like we're coming late to the party, but, there's no way you could acquire assets like Fortress's now, given today's commodity prices."

"I think the prospects for this particular company are very good. I've spoken to several base metal analysts—and this is prior to restructuring the company—I told them, assume there's no money in the bank, the company is in debt, and has no management. Basically, it was part-time for me; I did come in and clean it up, but it was my intention not to stay on as President and CEO. So, I told them to discount that. I just told them about the Poplar and the Davidson and gave them some quick numbers without all the details. They didn't realise there was a full feasibility study done or anything like that, and they felt it would justify a minimum of $50 million market cap, to as much as $100 million.

"Obviously, the growth from there, as we build a management team, and become well funded, that will all help. And today, we're probably at about a $10 million market cap. So, we feel from a share perspective; the company has a lot of upside potential.

"And, in terms of the metal markets, we see both moly and copper being very strong for the years ahead, with the growth that's occurring in China and India. It's not like the gold rallies that we've had historically. This is really an industrial type of commodity and it's based on consumption. We don't see that consumption slowing down. Like, uranium right now, I feel is ahead of itself; there's some emotion involved. But, we don't see that in the base metal market."

Ultimately, with the potential for an even rosier outlook, Fortress Base Metals' successful restructure and rejuvenation means that they can concentrate on delivering what's best for their shareholders. "Absolutely," says Wasilenkoff. "Now we can build off of this; we've got everything in place. We've just completed a small financing, and we're ready to build the geological team and advance all three properties."

Despite his job being almost done, he admits, "I see myself staying on the board of directors for several more years, until these things get into production because, I know I can assist. "I won't be the driving force behind Fortress—to build a successful company, you need somebody to put their blood, sweat and tears into it, to work 18 hours a day, 7 days a week. I won't be doing that for this company, but I'll be there to assist that management team," Chad Wasilenkoff says.

A modest contributor, without any sense of ego or need to be credited with his recent achievements, Wasilenkoff seems set to fade back into the scenery. "I'm just a very interested shareholder, and board member," he says modestly. "As I said, I usually start as the President and CEO, and Director, and then build up a management team with industry-specific experience. And then, usually, I hand the reigns off, or the day-to-day operations off, to these particular experts."

He adds, "We've just signed up a new President and CEO at Fortress, whose name is Damien Reynolds. He's run several successful mining companies." Reynolds was the Executive Chairman of Tournigan Gold Corporation, and Founder of Kyoto Capital Partners and Kyoto Funds Limited. "Damien is very good at what he does—building and taking these companies to that next level," says Wasilenkoff. "He's built a couple of other big success stories on the Toronto Stock Exchange."

"Arni Johannson has joined the board of directors," Wasilenkoff notes. "He was also on the board of Titan Uranium." Johannson has 15 years public market experience and is a Director of publicly-listed Mega Uranium Ltd.

"And then we've got a very strong management team under that with geologists and accountants; personnel like that. We're just finalizing our new geological team. We wanted to attract some of the best people in the world. Naturally, it was difficult for the company to do that, say, six or nine months ago, when it had a poor share structure, it was in debt, and all the properties needed to be reworked in terms of getting them into good standing. Once we'd put all those pieces back together, and then also completed a financing, now we've got several million dollars in the bank.

That means, a new potential candidate on the geological side could see the opportunity here—we're well financed, we've got the proper structure in place.

"Of course, at this point in time everybody who's half decent in this industry is gainfully employed," points out Wasilenkoff. "But, really what attracts people is not the salary but the option package"—and how far they think the stock can go as the company develops its properties. "Now when somebody comes in and evaluates that, I think they'll be comfortable and confident that their option package will be worth something down the road."

And geologists are coming on board. It is, after all, a prime time for moly—an asset that Fortress Base Metals is hoping will put the company on top in the years to come.

With their diligent restructuring behind them, and a trio of promising projects in their pocket, it seems that Fortress is finally working on that last piece of the puzzle. "Yes, we are just putting the geological team in place; they just provided the executive summary of the sort of work program they would like to do. So, right now, we're ready to proceed; to put some work into these three great properties that we have."

GEODEX MINERALS LTD.

With a monopoly on polymetallic properties in New Brunswick, Geodex has honed in on some significant untapped resources, such as Sisson Brook, a large open-pittable tungsten-molybdenum deposit.

Vice President of Exploration and company Director Jack M. Marr (M. Sc., P.Geo.) explains their strategy, in his affable, native Scottish brogue. "The whole area in southwestern New Brunswick, it's really a package of metals—tin, tungsten, moly, and indium. We've probably got 20 properties there now, and you tend to get a different metal coming out on each one," Marr says. "We've been building up a property inventory at a great rate. And I think we've got some of the best properties in that region."

The fact that the Canadian Maritime province of New Brunswick has been mostly overlooked by the mining industry in Canada is a double-edged sword for Geodex, On the one hand, the distinct lack of competition when it comes to staking claims is a great advantage, but on the other hand, the ignorance of stock brokers, especially on the West Coast where Geodex is based, makes promoting New Brunswick properties a challenge. "New Brunswick has been neglected," notes Jack Marr. "It's very lightly staked and lightly prospected. It's just not that well known to companies, whereas Newfoundland, by comparison, is a lot busier. Brokers in Vancouver know where Newfoundland is; it kind of hangs off the end there. But, you try to get them to put their finger on New Brunswick and it's not so easy."

The multi-metallic characteristics of New Brunswick, however, make it a desirable location for exploration and development. "Tin, tungsten and molybdenum; that package is pretty well confined to the East coast. There is obviously some tungsten in the Yukon but not that much in British Columbia," says Marr, a 35-year veteran of the industry. "If you look at the Maritimes in Canada, they're particularly enriched in those metals," a fact that Marr attributes to continental drift. "At one time, before the Atlantic opened up, the Maritimes were connected to Cornwall (UK) and a lot of the mines in central Germany. So, there is a belt there that was in existence before the present Atlantic Ocean started to open up. It's the same with tin. There's almost no tin in British Columbia. The only tin mines in Canada are in Nova Scotia, places like East Kemptville. It's just a very rich source of these elements, and you could link it up with the old tin deposits in Cornwall, and central Europe—same age, same kind of stuff. Therefore, we're pretty convinced there's a solid belt there."

That conviction is driving Geodex to concentrate its efforts wholly on New Brunswick. "There may be other companies in New Brunswick, working on gold and base metals; but we've definitely got the market cornered for tin, tungsten, molybdenum and indium," Marr concludes.

At his desk in Geodex's head office in Vancouver, B.C., the 61-year-old Marr strokes his bearded chin, thoughtfully. "I've worked on just about everything in my time; 'everything, everywhere' it would be true to say - even diamonds. To a geologist, the process is the same. It's like a chef baking a cake. Each one may require different ingredients but the same chef is perfectly capable of making different cakes. Geologists are like that, too."

A skilled geologist, Marr graduated with an honours degree in Geology in 1968, from the University of St. Andrews in Scotland. "I came over to Canada to work," he says. "I told my mother-in-law that I'd take her daughter to Canada for a few years, and never went back." Marr received his Masters degree from the University of Manitoba in 1970.

"I have actually worked with molybdenum in British Columbia in the past," Marr explains, "I was there during the first wave of exploration in the late 1960s and early 1970s. So, I know the porphyry copper business." Marr went on to work with Esso Minerals Canada in Vancouver for eighteen years, from 1972 to 1989, becoming a district supervisor in 1985. His experience with Esso included exploration work in B.C., the Yukon Territories and the Western Shield, plus two years in Western Australia, assisting in the discovery of several major deposits.

Moreover, during his time at Esso, Marr met Geodex's future President Jack M. Maris. In addition to having similar-sounding, near identical names, the pair hit it off. "Jack optioned a property to Esso Minerals and we got to know each other," says Marr. "Esso Minerals went out of business shortly afterwards and I went to work for Kennecott Canada Inc. I was Kennecott's regional exploration manager in B.C. for two years (January 1991–September 1992). I did a lot of work on diamonds at that time. Then I was doing some consulting work in 1993. Jack Maris had inherited Geodex and he wanted to get involved in the business, and he asked me if I'd come along. We've been partners in the group ever since."

Geodex, in all its incarnations, has been in business for about half a century. "There were a variety of names since the 1950s, I believe," says Marr. "It was called Kamad Silver at one point, because they were mainly doing a lot of work up in Kamloops. When Jack and I inherited the company, the name was changed to Agate Bay Resources. Years later, it became Geodex.

"The current Geodex team has been around since 1993 and, like many junior companies, we've worked in Mexico, South America, Eastern Europe and across Canada. We've paid our dues. And when you see problems in South America—nationalism raising its ugly head in Venezuela—you realize there are good reasons to be in Canada. The culture is so much easier to get along with," Marr says, speaking from experience.

Now Geodex is setting its sights on the stable political environment of New Brunswick and, in particular, on its Sisson Brook advanced-stage tungsten-moly-copper deposit in the central belt of the province. "Most of the moly we've found so far is at Sisson Brook, just north of Fredericton," Marr explains. With inferred resources of 167 million tonnes - 77 million pounds of molybdenum at 0.021% and 217 million pounds tungsten at 0.059%—the Sisson Brook is Geodex's flagship property. "We're pushing the Sisson Brook along towards development," says Marr. "We'll have a scoping study completed by the fall of 2007, produced by Wardrop Engineering. And we'll be doing metallurgical test work and an environmental base-line study, also in 2007. Hopefully, we can go straight from the scoping study to a pre-feasibility study. Then we hope to go straight from pre-feasibility to feasibility, and then start construction a year after that. So, I think we'll be able to advance this fairly quickly."

Geodex has budgeted $5 million for Sisson Brook's pre-feasibility study. The study will be based on an open pit mining model with a predicted 20-year lifespan. So far, as Marr explains, things are looking good. "We're looking at 20,000 tonnes a day for 20 years—that would require 140 million tonnes. This last report gave us 167 million tonnes, depending on the cut-off used. So, this first year of work we've really got all the tonnage we need. But, what we'll do now is look for grade and try to improve the grade and maintain the tonnage. As we keep drilling, we'll add higher grade tonnage and toss out the lower grade tonnage. This way, we can improve the grade of the overall project and make the economics more interesting. If Sisson Brook goes into production, that's dollars a share for us; no question." Marr continues, "We really made a huge advance last year. Basically, we came up with all the tonnage in the first phase of drilling, which is almost unheard of."

Marr anticipates a lot of support for getting Sisson Brook up and running. "You know, the Bathurst mine (the world's largest underground zinc mine) has really been the main mine in New Brunswick for the last 50 years, but it's only got a few more years of ore left. It's just about ready to expire, and is expected to close in 2009. So the government will be looking for something to replace that old mining camp. Therefore, I think we'll get all the help we need from New Brunswick's government to put Sisson Brook into production."

President and Managing Director of Geodex Minerals, Jack Maris says, "It's great because we're ahead of schedule and, yes, we are getting tremendous support from the New Brunswick government. I was there in June and met with the Minister of Natural Resources, Hon. Donald Arseneault. He met with us for a breakfast meeting with a bunch of analysts.

"It really impressed everyone that the minister would take time out of his busy schedule to attend our breakfast meeting, which was strictly for Geodex and the analysts. These kinds of things really are very helpful.

"This just proves the government's sincerity in being involved in the development of the Sisson Brook project. They realize the potential for employment and various other things, of course. But, New Brunswick has lost a lot of people to Fort McMurray (Alberta); now these people will hopefully be coming back with more projects coming on stream.

"Overall," says Maris, "I'm just very pleased with the way Sisson Brook is developing."

As Maris looks positively towards the future, Vice President of Exploration Jack Marr keenly describes Sisson Brook's past history: "The deposit wasn't found until it was partially drilled by Kidd Creek Mines in the late 1970s, early 1980s," explains Marr. "One important point about New Brunswick, and the same applies to Newfoundland, is that there isn't really a lot of rock exposed. When the glaciers left, 15,000 years ago, they plastered this layer of glacial till and overburden on top of the bedrock. So, that's one of the reasons why Sisson Brook wasn't discovered before. It really doesn't outcrop. You can log it, skidoo over it in winter time, but you've no idea what's down there. And the deposit doesn't respond to airborne surveys. Our Sisson Brook deposit is what we call a disseminated deposit. The sulphide content is fairly low. It was flown several times, by companies looking for a Bathurst-type deposit—they are massive sulphides; really heavy sulphides. But, it didn't respond to airborne surveys, so nobody found the Sisson Brook deposit until much later."

Kidd Creek's drilling had found three substantial areas of mineralization along a two-kilometre stretch; but when metal prices slumped they gave up the property. The two northernmost mineralizations on this stretch consist of two copper-tungsten zones. Five hundred metres to the south lies the third mineralized area, an apparently larger tungsten-molybdenum zone. Based on Kidd Creek's results, all three zones indicated enough size and grade for a cumulative tonnage of about 35 million tonnes, though none of these findings are compatible with modern NI 43-101 regulatory standards. Still, these findings were significant enough to intrigue Geodex.

Currently Geodex is earning a 70% position in this major copper-tungsten-molybdenum porphyry system, with a view to going into production in 2010 ("probably about three years, give or take," according to Marr). A program of survey work undertaken in 2005 and two subsequent drill programs have added to and extended the three mineralized zones. A 50-million tonne deposit was initially the target within range.

"Essentially, we're at the early development stages. We drilled the project last year; we know it's big. We've got millions of pounds of moly and tungsten there. It's very wide, about 300 to 400 metres wide in the middle. It lends itself very well to an open pit configuration, which gives us more flexibility - we can get more throughput, it's much cheaper to develop. And I think if you add that to all the other benefits in New Brunswick, Sisson Brook is looking very attractive."

Sisson Brook is agreeably situated, with rail and hydro throughout, plus a comprehensive network of logging roads. "Oh, it's absolutely amazing. You can get to the Sisson Brook, you can land at the Fredericton airport, and be on the site in an hour. What's more, there are no Aboriginal claim problems, it's all Crown land; no access problems; we've got a power-line right across the property; water; a road system; small towns around, so you can find labour and contractors. When I give presentations on it, I call it a 'dream location' - and I'm a geologist, I'm not a promoter."

Marr explains, "If you compare it to the problems that companies have in British Columbia, you know, there are some real challenges there; you're up in the mountains, you've got weather problems, you've got no road access. Most of the properties in B.C. have no power to them. If you look at 90% of the projects in B.C., the advanced projects, they are probably being held up by the need for power lines and access roads or Aboriginal claims or something like that. Many environmental problems, too. Then, once you've built your mine, you've got to figure out how to get your concentrate to port. I just don't think that point is understood.

"Compared to that, New Brunswick is a dream. The difference is incredible. I don't think that has gotten through to a lot of people yet. Of course, here you're close to ports. And we even have a rail line and a provincial highway crossing the property at the north end. I mean, it's a fantastic location," he says.

President of Geodex, Jack Maris agrees, "You know, the logistics are indeed absolutely incredible at Sisson Brook. The fact we have the railway right beside it, we're two hours from the main port Saint John. If you handpicked a deposit anywhere in Canada, you wouldn't find a better place than where Sisson Brook is located right now."

The peak season for working in New Brunswick - as elsewhere in Canada - is summer. "We don't do much work between January and March. However, we could. We actually drilled through that period this year. You could do it. There's a lot of snow and cold weather, but nothing compared to the shield or the Northwest Territories." Geodex has a year-round field office in Fredericton with a field staff of "this year, up to eighteen people and quite a number of geologists. We'll have three drills working for us this summer," says Marr.

Although busy with other aspects of running the company, Marr still likes to dabble in the geological aspects of the property. "Oh yeah, I do. I was out there last week, in fact. But, most of my work now is management. Making sure that the money is spent where it should be. Just to work on the progress of the properties. Typical management stuff—but I still like to see the properties, advise on the geology, and look at new properties," Marr says. "The experience is there. And it's difficult to teach this stuff in school."

Finding locally-based employees is proving relatively simple for Geodex's New Brunswick office. "Like we said, we've got a government that is willing to help us as much as possible, because half the labour force is in Fort McMurray - and like everybody, these guys would rather work at home. However, New Brunswick needs jobs. And Sisson Brook is located in a rural part of the province. We'll be making work in the hinterland there, but if we're shipping concentrate back to the port it's going to make work there as well. So, the effect will multiply to the whole economy. And I'm sure the government is pretty aware of that. So, ultimately, we'd have no trouble getting a workforce there if we decide to build a mine," says Marr.

Sisson Brook isn't the only feather in Geodex's cap, though. "We have the two areas," notes Jack Marr, "the Sisson Brook, and then we have a package of ten or twelve properties around the old Mount Pleasant mine that we're exploring for tungsten, moly, zinc and indium; a variety of elements. I should point out that Geodex does not own the former mine property. We just made a deal with Teck Cominco which is interested in those properties. They like the area and they'd like to increase their indium output. They produce about 50 tonnes at the moment from the Red Dog mine in Alaska and they would like to boost that. New Brunswick is a rich source of indium."

Geodex refers to the entire project as Mount Pleasant West, in reference to the less-explored prospects within a 20-kilometre long belt to the west of the mine. With features such as soil geochemical anomalies and tin- and indium-bearing boulders, indications are that something better may be found at depth—something currently obscured by the thin mantle of glacial till.

Indium is a rare, soft, and malleable metal, used in liquid crystal displays, such as those seen in the widely-popular flat-screen televisions. As a result of this industry the price of indium has increased from $70 (USD) per kilo in 2001 to over $1000 (USD) per kilo today.

"After the first year or two of exploring at least a dozen properties there, Teck will be funding us to work on it. We'll just keep going, and we have a major partner and a source of funding. We're pretty optimistic as things go along that it's going to work out," says Marr.

Just like Sisson Brook, the Mount Pleasant area has this secret trove lurking, unseen, beneath the surface. "People, I think, are really unaware of it. Particularly in the Mount Pleasant mine, it's significantly rich. In fact, it's still the best source of indium in Canada." Due to their ownership of the Mount Pleasant camp (8,640 hectares, or almost 90 square kilometres, in southwestern New Brunswick, surrounding the previously operated Mount Pleasant mine), Geodex is considered the world's largest land holder of premium indium exploration properties.

On Geodex's Mount Pleasant West properties, a 2,500-metre drill program was recently completed on five target areas. Significant results from the first four of nine drill holes revealed 31.7 metres of 62.81 g/t indium and 2.8% zinc. Back in June, the company announced 10.23 metres of 128.64 g/t indium and 3.45% zinc.

From 1983 to1985, Billiton Exploration Canada Ltd. operated a fully-permitted mine at Mount Pleasant, including hydro and tailings facilities. The mine, now owned by Adex Mining Inc. produced for a short period at a design rate of 650,000 tonnes per year until decreasing metal prices made the operation uneconomic. The Billiton mine, however, was based on reserves of tungsten and moly and predates the discovery of underground tin and indium deposits within the mine. It is now suggested that Mount Pleasant may be the largest high-grade indium resource on the planet.

The easy logistics of the Mount Pleasant West properties, combined with an abundance of prospects and a nearby mining plant, make this an attractive program for Geodex. They continue to target numerous showings, float boulders and soil geochemical anomalies on the property for tin, tungsten, molybdenum and indium.

"And we're still taking on properties," Marr states. "In June, we acquired four new properties in southwestern New Brunswick, with indium and uranium potential; all around the Mount Pleasant area." Marr always has an eye open, "because no one else is doing it." He says, "One is a cobalt property. Also, we've staked two properties for uranium, but the area is so polymetallic that we're finding moly and tin on those claims as well.

"The Mount Pleasant mine in New Brunswick is probably the most polymetallic mine in the world, with the most elements. The whole periodic table is in there!" he smiles. "As a geologist, I've never seen anything like it. It's amazing. Likewise, we go into a cobalt property and find tungsten and tin, on a uranium property we find indium, and so on..."

Much lower down on Geodex's 'to do' list is another project called Harry Brook, which was acquired in October 2004 by staking three properties along a 10-kilometre section of the gold-enriched Annidale belt. "Oh, it's very early stage," says Marr. "It's just a boulder, basically, with high grade gold numbers in the Annidale gold belt; slightly to the east of all our tin-tungsten."

The Annidale belt is a narrow, laterally-extensive belt of rocks in south-central New Brunswick. It was explored by various companies in the 1980s. "The interesting point about this boulder is that it has the highest gold assay ever recorded in New Brunswick. It was actually sampled by a government geologist—an independent survey, really," Adds Marr. The Department of Natural Resources announced that a geologist in their group, Susan Johnson, had made the gold discovery at Harry Brook. Johnson had obtained an assay of 1320 g/t gold (nearly 40 ounces per tonne of gold) from a boulder in to the northeast along the same structural trend as Geodex's Harry Brook property. This obviously suggests the possibility of high grade gold deposits in the region.

"Yes, so the gold was running about 40 ounces per ton, which is just unheard of at $600-plus an ounce," laughs Marr. "Unfortunately, we haven't managed yet to establish the source of the boulder. We've been doing some work there, but not at a high level. We've been really too involved in the moly, the tin and the tungsten. But, you know, we're working away there every year. We'll see how it goes."

To manage a plate full of prospects, from Sisson Brook's tungsten and moly to Mount Pleasant's tin, indium, and more, and even Harry Brook's golden potential, Geodex's Vice President of Exploration is supported by a well balanced and experienced team. "We have a very good group, who work well together, from top to bottom," Marr says.

President and Managing Director of Geodex Minerals Ltd., Jack Maris is principally responsible for the administrative and management affairs of the company. From 1985 to 1998, Maris was President and Managing Director of both Kamad Silver and Agate Bay Resources Ltd. Under his direction, those companies were reorganized and negotiated several option/ joint-venture agreements with major mining companies. Prior to 1985, Maris was an independent businessman involved in the construction and food industries.

Jack Maris works closely with Jack Marr, and Geodex's Vice President of Business Development Christopher Anderson. "I think Chris deserves a tremendous plug on this, as well," says Maris. "He's really gotten the story out to the analysts and to the brokers. Chris also organized our last financing, with Jennings Capital Inc. We raised a total of $10.5 million dollars; about $7 million of that came from Jennings. Chris was very instrumental in getting that completed."

With a successful career as an entrepreneur in all phases of business, with an emphasis on strategic planning, communications, and creative marketing; Chris Anderson has developed and published several news magazines/papers still in circulation today, created an Internet service provider company, and started and operated several retail operations. Anderson was the financier of all his own projects.

"Basically, here at Geodex, I'm responsible for coordinating the financing and developing the story; working on the marketing side, strategic planning, that kind of thing," Chris Anderson says.

"What we've done is integrated with the community; we've built good human resources there. As far as the company goes, we're full of geologists. I'm one of the few in the company without a background in geology," he admits. "But, I've worked in marketing, and when it comes to developing the story, you know, New Brunswick is the only place in the world where you can mine indium, stand alone, and actively focus on that. When you say that, and that the Mount Pleasant camp is the most polymetallic deposit anywhere probably in the world. Well, essentially, we chose one word and one commodity to talk about - indium. Not only because it was in high demand, but because if you say indium is used for flat screen TVs and monitors; people instantly understand."

Anderson concludes, "That was one of my key roles; pulling that information out and bringing it to the forefront." A dedicated and focused individual, he says that he ultimately follows a motto for success: "'Plan your work, and work your plan – that way, you can conceive it, believe it, achieve it' - I think the failing of a lot of public companies is that they don't do that."

Chris adds, "Of course I wouldn't be able to do my job effectively if I didn't have the help of companies like Renmark Financial and Scott F. Gibson & Company. These companies and the relationships we have developed with their staff have been key in helping us achieve our objectives. Developing the right team has been instrumental to our success and I think we have managed that on all fronts."

That includes J. John Jardine, Geodex's Chief Financial Officer. A professional accountant, he graduated with a Bachelor of Commerce degree from Carleton University in Ottawa in 1975 and earned a C.M.A. in 1978. He has 14 years experience working as Chief Financial Officer and Director with mineral exploration companies listed on the TSX and ASE exchanges.

"There's also Jack Patterson, who is one of our directors. He used to be the Managing Director of the Chamber of Mines here in B.C. He lives in New Brunswick and helps us with research; that really helps, too," says Jack Marr. Patterson is a Geology graduate of the University of New Brunswick. He also attended graduate studies at the University of Manitoba. From 1962 to 1971 he was employed by a major mining company exploring for base and precious metals throughout Canada. From 1980 to 1998—when he retired and moved to New Brunswick—Patterson was the Managing Director of the British Columbia and Yukon Chamber of Mines. He became a Director of Geodex Minerals in 2003.

Additionally, Neil Humphreys is Chief Geologist for Geodex. He received a Master of Science in Mineral Exploration from Queens University in Ontario in 1981. He is a professional geologist with over 20 years experience in project evaluation and development in Canada, the U.S.A., the South Pacific and Latin America. During his career Humphreys has worked with companies including BP Minerals, Esso Minerals and Teck Corp. He is presently in charge of coming up with new projects in and around our present operations.

Dave Martin, B.Sc., P.Geo., CGA, has twelve years of mineral exploration experience, primarily in New Brunswick. He is presently the Regional Geologist and District Manager for Geodex Minerals. Martin was an Exploration Geologist with Lac Minerals, Noranda, Inco, Novagold, Acadia Mineral Ventures, and Adex Minerals. Moreover, Martin played a key role in the discovery of gold occurrences in southern New Brunswick at the Armstrong Brook and Annidale properties. Plus, while with Adex Minerals, he was involved in the discovery and modeling of tin-indium resources at Mount Pleasant.

Dr. Sheila Watters (Ph.D., P.Geo., FGAC) is a New Brunswick resident and Senior Geologist for Geodex. She received her B.Sc. in Geology from Acadia University and her Ph.D. from the University of Western Ontario. Her 25 years in mineral exploration include projects for both major and junior exploration companies with a variety of mineral commodities, working mainly in Canada but also abroad. Consulting since 1990, Watters has conducted field projects for the exploration industry as well as ore deposit research projects for the New Brunswick government.

"I know I'm biased, but I think we're a very solid group," notes Marr, philosophically. "I think we present ourselves as a very technically competent team; I think everyone would agree about that. We've been beefing up our Investor Relations in the last year, trying to get the story out a little better – that's where Chris Anderson has been involved. And, the public are becoming aware of us. We really have some very solid assets.

"And we're a well established group. Four of us have been working here for the last five years. We've managed to hire very good people. Our group in Fredericton has doubled in size every year for the last three years. The budget has doubled, too. All our geologists and technicians are in New Brunswick and you couldn't get a better quality of people. Most of them actually live at home and just travel to the deposit every day."

New Brunswick is clearly where Geodex's heart is. "This is something that we've been aggressively pursuing, because we enjoy working with the people of New Brunswick, and with the prospectors as well as the government there," concludes Jack Maris. "It's just been a real home for us, and I'm really pleased with the way it's all advancing."

Jack Marr adds, "Yes, I think we have a good focus on New Brunswick now, with lots of good contacts. Our focus is on the moly-tungsten-tin suite." In addition, "well, the demand for indium, as a high tech metal, is just sky-rocketing. You know how the market has gone in flat screen TVs. Supply is falling well below demand," suggests Marr. "But, I think the huge interest in the future will be solar cells; improving the efficiencies of solar cells."

Marr continues, "I really call indium 'the wonder metal'. There's going to be a lot more of it used. And then moly, I think, is setting new standards - being added to stainless steel, and some of it is replacing nickel.

"Actually, I think moly's main use eventually will be in high strength steels and corrosive resistant steels. The high strength steel market is interesting because there's about 1.6 billion tonnes of ordinary carbon steel produced every year. If you add a little bit of moly to that steel, you can produce a much higher strength product. Places like the auto industry are really interested in this because it means they can use less steel, which means the car is lighter, which means you get better gas mileage.

"Also, a lot of the uses now for steel are related to corrosion resistance. Pipelines, nuclear power plants, desalination plants; and moly is the element of choice for those."

Of course, tungsten is the other big metal at Geodex's Sisson Brook. "We're pretty optimistic about tungsten. The Chinese have always flooded the world markets with tungsten and now that's stopped, and they're actually importing tungsten. It has the highest melting point of any metal. It's good for hard-grinding tools, drills and drill bits. And it's even being used for wedding rings and watches. Like titanium, it's getting to be cosmetic.

"If we keep up the pace of industrialization, there's always going to be demand for cutting steels and high strength steels. Tungsten might surprise everybody and there's much less tungsten around. It's usually found in higher grade underground deposits. There isn't really much tungsten in open pit deposits. So, we've got all the benefits of scale there on Sisson Brook."

Indeed, Marr points out that, at Sisson Brook, the "metal in the ground at current prices is worth about $5 billion. We have 70%, that's $3.5 billion. If you take 10% of that as being Geodex's profit margin—you know, we'd like more than that—that's $350 million, divided by 50; that's seven bucks a share. We're pretty convinced when we put this all together, we'll be in the money. We'll get the scoping study by the fall, which will confirm the whole thing, and then we'll be on from there to pre-feasibility and feasibility," he says. "And that's only one of the properties—that doesn't include all the work we're doing at Mount Pleasant and so on."

Furthermore, Marr says, Geodex will keep diligently acquiring and growing its inventory of New Brunswick-based properties. "We have two or three more on the hook at the moment. Because, simply, we have no competition. At the Sisson Brook project, we've staked about 400 claims; not a single claim has been staked anywhere around us. We've monopolized the whole area.

Evidently, as the 'key' company in the province, as far as juniors involved in tungsten, moly and indium; Geodex's monopoly looks like an unrivalled success.

GOLDEN PHOENIX MINERALS

There is a framed photograph gracing the entry hall of Golden Phoenix Minerals headquarters, outside Reno, Nevada. The photo features Earl Harrison, a former Golden Phoenix employee and its first Mine Manager assigned to a remarkable molybdenum project located in the northwest corner of Nevada, near the tiny outpost of Denio. In the photo, Earl - a burly and bearded 40-year mining veteran – is seen standing 300 feet underground in front of a massive vein of molybdenite, some 10 feet wide, cutting a dark gray swath across the background of host rock. Earl often said of that vein, called "the Sylvia," that a man could mine for a lifetime and never have a chance to work a virgin vein of such high grade. It was Earl who proudly led the first mining crews 2000 feet into an old decline to mine the Sylvia vein, and thereby establish the newest primary moly mine to come on line in nearly a generation. The name of the mine is Ashdown, and it boasts molybdenite grading at current values equivalent to between two and 40 ounces of gold per ton. Grades like that are what the old timers called "bonanza."

Under the guidance and funding of Golden Phoenix, Ashdown Mine shipped its first supersacks of molybdenum concentrates in January of 2007, and since then, has been steadily ramping production on what is considered one of the richest moly deposits in the world. Its unique mineralization is concentrated in a narrow deposit that stretches across 1200 feet of strike, open on both ends. The vein lies in the ground at a 45 degree angle, like a giant serving tray, tilted underground. As of this writing, men and equipment are gnawing away at the vein about 350 feet below the surface, and hauling as much as 100 tons of rich ore daily to a specially built mill two miles away. There the rock is crushed, ground to a powder in a ball mill, and the molybdenite is separated through a series of flotation cells, with the final product ending up dried and sacked for shipment. At current market prices, each shipment, consisting of 12 supersacks containing a combined total of about 60,000 pounds of concentrates, is valued at between $700,000 and $800,000.

David A. Caldwell, Golden Phoenix' Chief Executive Officer, has helped nurse this one-of-a-kind mine to life over the past several years, and is quick to point out that even though it's current milling capacity is very small by typical moly mine standards, its grade more than makes up for its size. Caldwell feels that Golden Phoenix has extraordinary potential specifically because of the grade seen coming out of Ashdown. "We look small, because people tend not to see past the face of it. Yet we have about 100 times the grade of the average moly mine. We have what I call quality pounds, with grades in the initial stopes around 3% Mo (elemental molybdenum), and life-of-mine numbers well in excess of 1%. This means its 100-ton mill can produce something between $100,000 and $150,000 dollars a day in product, and has an earning capacity equivalent to a 10,000 ton-a-day mill at a moly operation with average grades. Grade is king in our business. Ashdown provides a graphic example of the core truth behind this old adage."

Dave Caldwell's enthusiasm over the potential of Golden Phoenix's Ashdown property, co-owned with its minority partner, Win-Eldrich Mines Ltd., comes from seeing major growth of the mine over the last few quarters. The fourth quarter of 2006 brought in $177,000. Then the first quarter of 2007 rocketed to $800,000. By the second quarter of 2007 that number had jumped over four-fold. The next few quarters will have production continuing to rise as the miners strive toward full capacity. "Reaching a sustained earnings level of $5 million in a single quarter is a corporate goal for the Ashdown Project, and a milestone where we expect people to sit up and take notice."

For Golden Phoenix, it wasn't always this way. In fact, as its name implies, the ten-year old public company has risen from the ashes of its past performance more than once. The first few years of Golden Phoenix were tough Caldwell recounts, "so tough that management had to work for no pay." Of course, this was the late 1990s, a depressed time for gold, especially with the Bre-X effect, which hurt so many legitimate companies in the mining field.

At this point in the story, Dave Caldwell pauses to smile and offers a poignant comment: "I've been with this company as a director since its inception 10 years ago and have played an active roll in resurrecting it from financial distress twice. It is fitting we named the company Golden Phoenix, because we have always be able to rise from the ashes.

In late 2004, two key men appeared on the scene just as the three outside directors, including Caldwell, were coming to realize that there were serious problems with the Company again. Both were private businessmen with no experience in public companies, but with expertise in start-up and turn-around situations. Bringing their long time entrepreneurial success to the table, Rob Martin (today, its President), and his good friend of 35 years,

Ken Ripley (who served as its CEO for nearly two years at the beginning of the transition) saw potential where others did not, and offered to assist. With the help offered from Martin and Ripley, outside Directors Dave Caldwell, Jeff Tissier and Ron Parratt were in position to avoid "an orderly bankruptcy, turning the lights off, locking the door and walking away. All you can really do from that prognosis is move up with each little victory over the incredible odds against you."

Fortunately for them (and future shareholders), what they also possessed was a good, solid, reputable mine, secured with a letter of intent joining their firm with the decades-old Win-Eldrich. In Caldwell's words, "Win-Eldrich had the underlying land position through unpatented lode mining claims, and was looking for a joint-venture partner. We arranged a marriage." Caldwell, Martin, Ripley and Mine Manager Earl Harrison brought the knowledge, money, and technical expertise to develop this impressive piece of property.

Golden Phoenix, in its latest resurrection, now owns three intriguing mining properties in Nevada and Canada. With rock being turned into cash and the first net profits distributed in the second quarter of 2007, one can easily see that Golden Phoenix has made a strong return from earlier times. Its managers, its teams of mining and technical personnel, and many investors believe that the company has indeed risen again.

David Caldwell, like so many in the mining industry, had his own phoenix-like experiences long before he got involved with this promising young company. Now in his late 40s, he was born in St. Paul, Minnesota, earned dual degrees in geology and geophysics at the University of Minnesota, then ended up at the New Mexico Institute of Mining and Technology, earning his Master's in geology and geo-chemistry. His goal was not to discover mineral deposits and run mines, but rather to become a volcanologist, as seen in his choice of thesis—a study of the active volcano Mt. Erebus in Antarctica. But the reality of life hit home when he saw the posting for a position to work in southeastern New Mexico and West Texas, looking for sulfur. Within the first year, he had paid off his student loans and greatly enjoyed the work as a one-man exploration office. That was his introduction into the industry, and three years later he was transferred to Elko, Nevada to work as an exploration and development geologist in the gold industry.

Caldwell states, "Nevada has an incredible mineral endowment and is ripe with mining opportunity. It is host to the second richest gold province in the world behind South Africa's Witwatersrand Basin, and is now moving toward status as a significant supplier of moly to the world."

He worked with Gold Fields Mining Company through two attempted takeovers by outside mining groups and finally a successful one by a multinational manufacturing concern who eventually sold the US assets off to Santa Fe Pacific Gold Corporation in 1993. He stayed on with Santa Fe, until Homestake made a bid in late 1996 for a merge, only to be trumped by Newmont in early 1997.

Tired of the disruptions, lack of efficiencies and control of an industry that had been under almost constant consolidation for decades, he and other colleagues from Santa Fe looked at starting up their own junior mining company where they could engineer the takeovers and benefit more directly from their success. After Caldwell and friends attended the 1997 Round-Up in Vancouver, B.C. they quickly realized, "Yes, we could start a company, but how do you bring a dozen people into a start-up organization that doesn't have a portfolio of projects together yet. No one wanted to pay the overhead for salaries in a top heavy organization. But the dream didn't die there; the friends splintered into two companies each with very different paths." One of them became Nevada Pacific Gold, while the other became Golden Phoenix. Caldwell was one of the five founders of the former, and became an outside director at the onset for the latter. Nevada Pacific listed publicly in Canada and succeeded in building enough asset value in its exploration portfolio to be acquired as part of Rob McEwen's four-way mega merge through his U.S. Gold vehicle, while Golden Phoenix listed on the Nasdaq's OTC-BB in the United States and focused on advanced mine-stage projects.

The CEO of Golden Phoenix presently lives in Elko, Nevada. His town is four hours east of Reno, but he maintains an apartment in "The Biggest Little City in the World" where he has been living most of the time. There has been too much going on in the world of his firm and its three properties. He describes the situation he finds himself and his company in, as of the summer of 2007, with both humor and insight:

"Our mission here was to build this company, and it took a single-pointed focus to actually pull this salvage operation off. What we've done here is, in some ways, an amazing thing: we've permitted and built a mine with no initial funding, using people who were essentially volunteering their time, all starting from a point of near bankruptcy—by all odds we shouldn't have been able to do it, but we did.

"Our success is a testament to the strong will and moral fortitude of the whole group. Moving forward the potential is great because of what we have accomplished through the team of people that has grown from five three years ago to over 60 today. Nine out of ten outfits attempting commercial mine production fail even if they have sufficient capital to accomplish the task. We have managed to make the transition from exploration and development on a property, into commercial cash-positive production with a combination of business savvy and dogged determination.

"Most of the world will never really understand what has happened here, or that the Ripleys and Harrisons and Martins and Tissiers and Parratts are truly the unsung heroes of this war. Where most industries mark their success through measuring the redistribution of existing wealth, it is on the backs of these men that Golden Phoenix has created wealth for its shareholder base where there was none before we started. That is a truly remarkable condition."

In the spirit of earlier reincarnations, Golden Phoenix's top officers, including Caldwell and Martin, took no salary until commercial production began at the end of 2006, and the two remain on half-salary as of mid-2007, "No I'm not getting rich, and I've got a second mortgage to cover the time when there was no money for salaries at all," its CEO declares. He thinks back to when he was 40 years old and in the process of laying himself off at Nevada Pacific, eventually going for over 20 months without a salary, rolling roof trusses and pouring concrete to make ends meet (at $100 a day) while doing his part to keep both public companies alive while waiting for the industry to come back. The running joke at the time was that Home Depot was staffed with out-of-work geologists.

Now, Caldwell finds himself "making hay in the middle of the biggest commodity boom cycle in recent history, fundamentally driven by China and India." He insists that Golden Phoenix is "not your average blue-sky company, but a blue-sky exploration and development company with robust earnings. We're a proven producer, and through that production have removed the awkward front end risk that holds most companies back. We don't have to go to the market to raise money for discovery and expansion because we are generating our own money." Naturally, there are always variables that you must keep your eye on, the price of molybdenum being one of them. Caldwell uses a $20 a pound model, expecting in 2008, full production at 100 tons a day—"by that time it is quite likely that, all things held equal, we'll be moving toward 200 tons a day at Ashdown, recommissioning our primary gold asset to establish a second cash stream into the treasury, and advancing our moly property in Ontario with test milling and marketing."

Caldwell's comment points to another jewel in their crown that adds to the allure of Golden Phoenix. It is the Mineral Ridge gold mine at Silver Peak, in the Gold Field-Tonopah area of western Nevada. Ironically, this is the same property which brought Golden Phoenix's previous incarnation to its knees when it failed to produced in the way that former management had expected. Their website predicts success in its next life given the wisdom gained by past mistakes:

> *"Mineral Ridge is a formerly producing gold and silver mine located near Silver Peak, Nevada. Historically, the mine is on record as producing 680,000 ounces of gold, primarily from underground operations. It has both underground operations and open pits, and features a six-acre heap leach operation, ADR facility and high volume crusher plant. It is currently on a lay-up schedule while engineering work is being performed to streamline future operations."*

The reason Caldwell feels this second property located in Nevada is so valuable is because, as an outside director, he was involved in approving the purchase of Mineral Ridge out of bankruptcy at the fire-sale price of $225,000, and he is clearly proud of that action. "It was a brilliant move by former CEO Mike Fitzsimonds and Vice President Steve Craig to capitalize on the low cost opportunity at Mineral Ridge, as it was for them to secure the Ashdown joint venture." Aside from the gold resources on the property which weigh in at over a million ounces of past production and remaining mineralized material in the ground, the true value is that this is a fully permitted, idle status gold mine in Nevada. Nevada has a track record of producing "elephants" in a stable political climate, both boding well for the investors in this Company.

The Company has decided to re-engineer Mineral Ridge by going back to a milling scenario that historically resulted in decades of consistent profits for previous operators with almost 700,000 ounces of gold produced. The rock there requires fine grinding to liberate the gold—that's a mill. By grinding you go from the 50% recovery realized by the three companies that have historically tried to heap leach the ore, to a recovery of over 90%. The costs of handling the ore are higher, but the certainty of recovery is much higher as well, so this approach effectively manages the inherent mining and processing risks much better.

Caldwell elaborates, "We're doing the third-party feasibility work that will lead us to a re-engineered mill that we'll build and have in production again soon. I'd like see the first gold bars—dore [mixed gold and silver] — to be poured within 18 months, which is aggressive, but possible if all goes as planned and we can find the materials to build the new processing infrastructure." This "jewel" is 100% owned by Golden Phoenix (subject to underlying royalties), but no other company has any claim on the expected fortunes that will derive from its development.

Golden Phoenix has a third property as well, this one located in Canada. It is named Northern Champion, combining both the mine's location and the firm's expectations. It is a molybdenum property located in the Ottawa Valley, near the nation's capital, but far from the usual places one looks for moly. It is interesting to report that this formerly-producing mine, now undergoing geological exploratory work to determine the scope of its resource and reserves, represents "the same thing that we're already mining down in Nevada; a moly deposit with exceptional grades, but this one is 100% owned by Golden Phoenix." It came into their hands as a submitted property by prospectors who were introduced by an existing shareholder, and, in a remarkably similar echo of its low-priced find with Mineral Ridge, the firm purchased it for $250,000 in Golden Phoenix stock in the summer of 2006.

The reason why the geologist/entrepreneur picked it up was that he was taken with reports and a 43-101 study that indicated a half-million pounds of high-grade moly drilled off and lying at the surface. Ashdown in Nevada has its riches about 350 feet below the surface, so the contrast is that "in Ontario, all we have to do is remove a layer of mineralized material that lies right at the surface, making the cost much less than that from the deeper veins in Nevada." By the end of 2007, Caldwell expects to produce a bulk sample for test marketing and metallurgical work in Ontario. There is still drilling to be done to determine how big the find may be, but he is confident that there are at least 10 to 15 million dollars of moly near the surface at today's prices based on the previous work.

"This deposit consists of large 2 inch wide hexagonal crystals of lustrous silvery-blue molybdenite. There are reports of some of the crystals reaching

12 inches in width with all of them possessing exceptional purity suggesting possible use in the lucrative aerospace and lubricant markets. That's why we call it Northern Champion."

Caldwell sits back with a contemplative look when asked what sets Golden Phoenix apart from its peers. "A lot of companies falter or fail despite having what seems to be the right stuff. Any company's stability can be envisioned using a three legged stool analogy. Often a company may have good properties, but may lack one of the other two legs of the stool—good technical people and access to money. You get all three legs firmly together, and you'll find a company that survives and prospers due to the inherent stability of the tripod support. A quick look at those that fail will reveal that they are missing one of those legs."

His argument to any investor looking at Golden Phoenix is that "we've demonstrated solid management, fruitful properties, and the ability to come to market and raise money. But more important, and quite different from a majority of junior mining companies, we're generating our own capital out of the ground. Our goal is to become a significant intermediate producer, and the current portfolio of producing and near-producing properties will provide the stepping stones needed to enter that rarified space."

Dave Caldwell has a voice filled with enthusiasm and charm. He talks happily about the four-dozen-plus employees working the mine and mill at Ashdown, with many of the mill crews coming off the local ranches, while the skilled underground miners come from all across America. It's not easy to find the skilled miners that the conditions at Ashdown demand, and the wage environment is extremely competitive in the industry today. Despite this, they are flowing in at a steady pace having heard about the once-in-a-lifetime opportunity that the "bonanza grade" of Ashdown affords.

Caldwell talks proudly, as well, about his board of directors and managers: Rob Martin, the President who, as an early shareholder, provided the Board with an alternative to bankruptcy, trading his life in Hawaii for 100-hour work weeks at headquarters; Ken Ripley, who came out of retirement in Seattle to take the position of CEO and then moved to the mine site to live in a camper, turning wrenches and pulling wire in sub-zero conditions because that was the only path to success in the darkest period of the salvage operation; Earl Harrison, the Mine Manager, who brought his equipment and underground mining experience to "build this thing out" while taking no time off during the first two years and allowing Golden

Phoenix to accrue payments for the use of his equipment, confident he would be paid in the future.

There is a solid technical team, including Don Prahl who has acted in top executive roles for Cyprus Minerals, Barrick Goldstrike and Cleveland Cliffs and is now Chief Operating Officer; Craig Patrick as Vice President Corporate Development with 39 years in the industry, starting as a miner while still in high school and eventually gaining his MBA while working for industry giants Cyprus Amax and Phelps Dodge; Ed Falk as a Senior Project Geologist who has over 30 years of industry experience with Kerr-McGee, Santa Fe Pacific Gold and Newmont; Wayne Colwell, son of a millwright, who grew up in remote Nevada mining camps learning along side dad, eventually gaining a BS degree in Chemistry and starting his own lab; the consulting business team of Dan Forbush (former CFO of Glamis Gold), Larry Kitchen (mine site accounting experience with both junior and major mining concerns) working with CFO Dennis Gauger (top end SEC/SOX compliance accounting) and Accounting Manager Lari Marlow. Lari, raised by her father, an underground miner, grew up in the Nevada mining camp of Mina, and has an absolute passion for the mining business. This team managed to keep it all going from an SEC filings and compliance perspective.

The Board of Directors hosts exceptional talent and experience with Jeff Tissier who has worked as an underground miner and gained degrees in Geology and Geoengineering from the Mackay School of Mines. He went on to become a CPA and was part of the team that packaged and sold the Getchell gold mine to Placer Dome for $1.2 billion dollars. Ron Parratt ran the Nevada exploration programs for Southern Pacific and Santa Fe Pacific Gold, Homestake and is a founder, President and CEO of AuEx, a Toronto traded junior exploration outfit that is successfully drilling gold on several properties. Both Tissier and Parratt were present through the turn-around phase of the salvage operation, and as part of the restructuring were subsequently joined by Kent Aveson and Corby Anderson. Kent is a Mining Engineer with extensive underground front line supervisory and mine management experience, and an expert in continuous improvement, currently holding this position at Barrick's Bald Mountain operation in Nevada. Corby gained practical milling and recovery experience at Sunshine Mining before gaining his PhD in Mineral Processing and Metallurgy and assuming the top position at Montana Tech's Center for Advanced Mineral and Metallurgical Processing.

This independent expertise ranging from exploration through operations, finance, mergers and acquisitions and metallurgy gives Golden Phoenix much greater depth than most junior mining companies.

"All of these folks are top-end people, with many decades of experience in all aspects of what is needed for success. As we gain credibility for being a producer, we have transitioned to attracting top mining talent, and having that talent attract further talent. The Company is growing organically in its capacity to achieve great things, and in my mind, that is the most exciting result of this turn-around." The CEO firmly believes that Golden Phoenix has a potential $1 billion-plus market cap as it moves to get its three mines in production while aggressively working to expand its known reserve base.

Simultaneously, other business opportunities are presenting themselves as Golden Phoenix fills the rarified space of the small-to-intermediate producer. The CEO is positive that the quality of its people and its track record of success give the company great leverage, which is why they have been approached with properties of increasingly high quality as operations have ramped up. "It's clear that we aren't on the radar screen of our competitors yet, but we have gained the attention of the independent property owners, which is a great place to be, from a strategic standpoint."

In the nearly three years that Dave Caldwell of Golden Phoenix has been part of the team driving the company from salvage to success, it's been nothing but all-consuming work, continuous 16 hour days. He joyfully describes his firm's retiring of over $6 million in debt off the liability side of the sheet in 2006, while settling law suits without having to take any money whatsoever out of their treasury or stock. What seems to please him the most is not his expectations of great wealth, but "putting all the pieces together, dealing with people and working to make something happen that wasn't otherwise going to happen." He sees a close link between the roles of the exploration geologist and the entrepreneur: both put together pieces of information, whether it's where to put the drill to make the discovery, or with whom to make the deal, the next joint venture, the hopeful acquisition.

"We are now molybdenum experts, we've got fully permitted gold in Nevada, and we know how to get the metal out of the ground and to market. We've become successful producers where most fail, and we have established legitimacy among our peers. Creating true value is never easy, but it's truly exciting to have been part of the team that made this happen and did the right thing by the shareholders. When I look at this objectively, using a fully diluted basis and the lowest and highest trading price for the stock, our management team has been successful in bringing value from a low of about $14 million dollars in the darkest days of the restructuring, to recent valuations of over $100 million and a sustained average of $80 million over the last several months. That is above average by anyone's standard, but the job is not done by any means."

The future for Golden Phoenix looks promising, indeed. It is wonderful to watch the "phoenix rising" myth become reality.

David E. Blann has plenty of experience in both 'the bush' and the boardroom. Speaking from his car phone, as he negotiates the twists and turns of B.C.'s mountain roads, the Happy Creek Minerals' President explains that he has amassed 20 years in the office as well as the great outdoors. As a Director and Vice President of several public companies and a qualified geological engineer (P.Eng); he is a professional in the exploration, development and mine operation industry.

"We find ourselves in a great climate for mineral exploration," Blann says. "It's no longer extreme suffering to get the funds to do some work."

Certainly, the jubilant-sounding company moniker seems apt, in consideration of this current climate. "The name 'Happy Creek' just kind of popped into my head; it sounded attractive. I think it's the only public mining company that's named after an emotion. Also, a lot of our new mineral discoveries—some really great showings and prospects—were made close to a creek," Blann notes. "It's catchy, too. People either love it or hate it. We've had mixed reviews on it, but most people do remember it; one way or another."

When creating Happy Creek, Blann - the adventurous outdoorsman – decided to combine his passions for mountaineering and geology with his interest in mining. "I have a personal motto," Blann laughs, as he travels en route to one of his regular on-site visits, "One foot in the boardroom and one in the bush." He likes to keep his finger firmly on the pulse, in touch with his investments.

With a degree in Geological Engineering from the Montana College of Mineral Science and Technology, and a diploma in Mining Engineering Technology from the B.C. Institute of Technology, Blann points out: "When I first received my degree, I ended up working at mines or advanced stage projects looking to go into production. It's good to know what a mine may look like and how it runs. That way, when you're out exploring and drilling holes, you have a clear idea of what you need to find to make it work."

Thus, Happy Creek Minerals' Blann states, "I had the mining and the geology backgrounds. I put those together and what I did was to take geology and try and translate it into something the mine engineers could understand. I worked to create my own niche as a go-between. So, I formulated this company to try and make the potential mining end of things as easy as possible."

The savvy, 46-year-old Vancouverite now carries the lion's share of the responsibility when it comes to running Happy Creek Minerals Ltd. "Because we're small, we have a small board of directors. I am currently the only full-time guy in the company," he notes. "But, all of us have ample experience in business. I've been traditionally an engineering and geological consultant. So, I took the opportunity to just say, 'I'd like to start my own company and do my own thing; do it my way.' Essentially, I have always wanted to find an opportunity in this business and go for it; and that's exactly what I'm doing."

Having a small team helps to keep costs down at Happy Creek. "It's quite a lean operation," says Blann. Although small, the company's strengths include management experienced in all phases of mining exploration, financing and public company management. The board includes Scott Smith, the CFO, who has a BA in Accounting and Business, and has both corporate management and business development expertise; Paul Reynolds (P.Geo), who also has 20 years experience including exploration and development, as well as 10 years experience operating and managing public companies; and David Ridley, who has a B.C. Certificate of Prospecting and has taken advanced level short courses, and has 20 plus years of dedicated prospecting experience in B.C., the Yukon and Alaska.

Happy Creek was incorporated in November 2004. "David Ridley and I were really the two guys that sat down and said, 'We personally have some great mineral properties, we'd like to see them explored further. Let's put these into a little company and try to attract some capital to explore them.' We were very successful as a private company." Consequently, Blann explains, "We attracted Canaccord Capital to carry out the Initial Public Offering; we signed up in February 2006 for that. By August 9th 2006, we had the IPO, and were thereby an operating public property." Happy Creek is listed on the TSX-Venture Exchange under the symbol HPY.

"We are really just over a year out of the gate as a trading public company," says Blann, "and in that time, we were successful in raising additional funds, and performing exploration on our properties. We feel that the best thing we can do for shareholders is continue our exploration and development on the properties to a point where they attract a major mining company's interest. And part of our overall strategy is 'diversified metals exploration'–that's been our slogan since 2004."

Happy Creek is engaged in the acquisition and exploration of mineral properties containing various metals such as copper, molybdenum, tungsten, gold and silver—all within the province of British Columbia. Their ultimate mission is to create wealth for shareholders through the discovery of economic mineral deposits, and then partnering or selling to larger mining companies who would develop and operate the mines. The Happy Creek strategy involves 100%-owned mineral properties that are close to existing and past-producing mines, and which have infrastructure in place to allow lower-cost exploration, development and production. In addition, their goal is to maintain a diversified metal portfolio in order to minimize risk and maximize value.

Despite their passion for diversity, moly is naturally becoming a predominant focus for the company. "Yes, we were cognizant of several properties having that potential all the way along," explains Blann. "We have actually explored these properties for over ten years privately and knew there was potential for molybdenum and tungsten. They were advanced by us with our own money and our own time, for a long time. Now it just happens that moly is one of the flavours of the day, and we've captured that. All our hard work and efforts and the risks that we took—they're paying off for us, it's going really well."

Happy Creek's Rateria property has turned heads because of its close proximity to an existing mine operation. "Rateria is a copper-moly project in the Highland Valley, and it's proving attractive to people because it's only 10 kilometres south of Canada's largest copper mine concentrator owned by Teck Cominco. In fact, the Rateria property is only six kilometres from the Highmont copper-moly deposit that Teck Cominco is re-opening for mining due to the high moly prices. Therefore, the obvious future buyer of our property, should we come up with a suitable deposit, is Teck Cominco," Blann speculates.

Rateria is comprised of seven mineral claims totaling approximately 3,832 hectares. "We drilled Rateria last year and we made a good start on a new discovery of copper, and we're encouraged by that. We followed up in early 2007, and drilling intersected 134 metres grading 0.16% copper, including 33.5 metres containing 0.31% copper from surface in a zone that is over 650 metres in length. Again, the real value of that property is that it's a new discovery and we're so close to the existing mine. If we come up with a good deposit, and we feel we are getting closer now, the opportunity is there to sell it to Teck Cominco. Because you don't have to build the entire mining infrastructure like a new mill/concentrator, even with modest grades, it could become a very valuable property."

In addition to the new copper zone, a substantial molybdenum soil anomaly was identified by Noranda in 1971. It covers approximately 1,000- by 500-metres in area, returning moly concentrations greater than 10, 20 and 40 ppm (parts per million). There are also several other historical prospects on the Rateria property dating from the 1950s to 1973. The property has been quiet since 1973, and these areas remain untested by drilling.

"Rateria is one of our main projects and the others are in B.C.'s South Cariboo region," says the Happy Creek President. The Cariboo properties consist of five groups of claims, known as the Fox, Silverboss, Hen, Art-DL and Hawk. Together, they comprise copper, molybdenum, tungsten and gold prospects.

Happy Creek has a 100% interest in the Cariboo properties, which were acquired in 2005. Located approximately 55 kilometres northeast of 100 Mile House, B.C., the Cariboo properties in part adjoin or are not far from Noranda's former Boss Mountain molybdenum mine that has production figures of 7.5 million tonnes at an average grade of 0.20% molybdenum (Mo), and a resource of 63 million tonnes of 0.074% Mo that is not to NI43-101 standards. The mine was closed in 1984 when molybdenum prices dropped to around $4.00/lb. Currently, moly prices are around $30.00/lb. The area around the Cariboo properties has been prospected extensively by David Ridley, one of the founders of Happy Creek, for about 20 years.

However, "The Cariboo properties are in the early stages. We haven't drilled those yet, but we are going to be pushing very hard to advance them this year," says Blann. "The Fox and Silverboss are the two molybdenum-tungsten bearing properties that are looking really good. They've come a long way in a short time. From the first glimmer, the first little showing with some molybdenum and tungsten; now we've started to get something on those properties that looks like they're both, potentially, very big discoveries. I expect we'll be spending a $1 million on these properties in 2007."

Like Rateria's prime location, the Cariboo claims are logistically advantageous. "The ones up at the Cariboo are largely leveraged by the presence of the former Boss Mountain moly mine," confirms Blann. Another of the Cariboo properties, the Hen property, located approximately 16 kilometres southeast of the Boss Mountain mine, is noteworthy for a low cost exploration program that identified significant lead, zinc, gold, copper and moly in soils and rocks over a substantial area that indicate a strongly mineralized system.

The 100% owned Hawk and Art-DL properties nicely round out Happy Creek's current portfolio. Hawk is a copper-gold-silver property. Historical drilling of the mineralized zone returned 0.61 metres containing 2.0% copper, 28.0 g/t silver and 6.13 g/t gold. One hundred metres south of that drill intersection, chip sampling of outcrops at the main showing returned 5 metres of 0.88% copper and 1.07 g/t gold and a further 600 metres south, 0.25 metres returned 2.17% copper, 15.0 g/t silver and up to 8.7 g/t gold. This property may be similar to Imperial Metals' Mount Polley Mine to the northwest. The Art-DL is a gold property with occurrences of 42.9 g/t gold over 1.0 metre at the DL adit quartz vein, and grab samples of 1.22 g/t gold and 620 ppm silver. This property appears similar to Skygold's Spanish Mountain gold prospect to the north.

The Silverboss property adjoins the distinguished Boss Mountain mine, formerly operated by Noranda Inc., and now owned by Xstrata PLC. In early 2004, Noranda dropped many of its mineral claims, retaining the mine leases that contain the former workings and tailings pond. The Silverboss property covers these former Noranda claims that are locally within 350 metres of an open pit.

A low-level airborne geophysical survey conducted over Silverboss, released by the B.C. Government in August 2006, identified comparable radiometric and magnetic signatures between the former Boss Mountain mine deposits and those occurring at lower elevations in the 10 Mile Creek area on the Silverboss property. This area is also within a few hundred metres of the same intrusion genetically associated with the Boss Mountain molybdenum deposits.

The geology, airborne geophysical data, and molybdenum values in silt and soil samples from 10 Mile Creek suggest a potential here for an undiscovered molybdenum deposit. In addition, Silverboss' Horse Trail zone contains molybdenite in quartz veins in a zone that extends westward from the existing Boss Mountain mine deposits.

Happy Creek's exploration of Silverboss in 2006 identified potential for new, large-scale molybdenum deposits around the previously mined area. Rock samples taken during this program returned results up to 0.637% molybdenum. Soil sampling revealed three areas containing coincident (overlapping) molybdenum, copper and tungsten, which supports the presence of underlying moly mineralization.

Well defined anomalies of molybdenum in soil covers areas approximately 200- by 700- metres, 100- by 350-metres, and 350- by 600-metres. Together there is potential for a large scale molybdenum porphyry system having overall dimensions of over 3 kilometres in length and 500 metres in width that has significantly increased the target size of the property. Happy Creek is continuing to explore the Silverboss property for extensions to existing molybdenum deposits and to locate new ones. Plus, extensive prospecting and soil sampling have identified areas returning significant gold and gold-silver values. Surface rock sampling has returned values up to 53.0 g/t gold.

With a view to the future, Blann admits that Happy Creek is giving serious consideration to acquiring the Boss Mountain mine itself, as the perfect complement to its Cariboo properties. "It has a lot of merit," he says. "The Boss Mountain mine was the highest average grade moly producer in Canada, and there's still a resource left there. They shut it down around 1985, when moly prices crashed. "It's a natural fit, as it's fairly obvious to anyone—why don't you acquire the old mine and put your two or three moly properties together? The whole is greater than the sum of the parts, and that's exactly what we're hoping to do here.

"In addition to having the Silverboss property around the Boss Mountain mine, there's the Fox—about 30 kilometres away," says Blann, "and, you could almost call that a 'twin' to the Boss Mountain mine. It's a brand new moly and tungsten prospect in British Columbia. We feel that it's extremely significant. We've got some tremendous moly as well as tungsten on surface, and over a large area." Happy Creek's exploration work has developed Fox into a significant new tungsten-molybdenum skarn prospect. The area known as the 'Discovery' zone, at Fox, was found to be comprised of molybdenite in skarn (zones of chemical replacement) and has returned trace to 4.99% molybdenum in grab samples. Soil geochemical surveys have identified anomalies approximately 1000 metres long by 50-200 metres wide and 800 metres long by 25-300 metres wide.

The presence of molybdenum and an intrusive rock with similarities to the Boss Mountain mine suggests promising potential for a similar metal source—only here contained in a skarn geological setting. Also at Fox is the 'Nightcrawler' tungsten zone that was discovered in May 2005 to the northeast of the Discovery molybdenum zone. Nightcrawler consists of a 1000- by 200-metre tungsten soil anomaly and surface rock samples up to 4.25% tungsten. Happy Creek will be trenching and drilling the Nightcrawler and Discovery molybdenum zones in 2007, along with follow-up soil and rock geochemistry on the north side of the granite stock.

Logging roads through the Fox claim, and hydropower lines approximately 15 kilometres west, are favourable economic factors. "And it's within trucking distance of the old Boss Mountain mine," highlights Blann. "Part of our strategy all along is to be in proximity to mines and infrastructure; we feel that's a real advantage in the long run, because once you have a deposit, it sure helps to lower the costs—because you have road access, and power already. That improves the probability that you can take it 'all the way.' And that was really what I wanted to do with this company. I've put properties in it that have the potential to go all the way."

Blann concludes, "If you can put those properties together, you've got yourself a scenario where you can really have a world-class tungsten and moly mine developed. In my view, it's pretty rare for a small company to come up with that kind of potential built right in, right out of the gate. Naturally, we're excited about it. We have the funds in place, and we are focusing the balance of 2007 exploration into the Cariboo properties and drilling them."

All in all, this cornucopia of metal properties is true to Happy Creek's enduring motto of 'diversified metals' and provides a breadth of opportunities for the small junior exploration company that isn't afraid of doing things differently from their competitors.

"You see, Happy Creek is a little different. I feel it is extremely important to be hands-on with the field data and to be out in the field, really seeing the rocks, understanding the geology, making conclusions and recommendations on behalf of the company that reflect the technical aspects of the geology," Blann says with enthusiasm. "It's important for management to truly understand geological process and mineral deposits, or one tends to lose touch with the reality of the prospects they are developing.

"It increases my workload, but I feel it's essential for me to have my feet in the bush, to be on the ground and actually see and understand things for myself. Shareholders deserve that. That's just my opinion. I know there's a lot of management and paperwork that has to be done; that's necessary, but I think it's really important—especially in a small company—that the guys who make the decisions have the best possible understanding of what's going on out there on their properties."

With a life-long interest in mountaineering and mountain climbing, he admits, "That's largely how I got into the geology and mining business. I heard that you could actually get a job and get paid to do stuff outdoors in the mountains. I thought, 'This could be something for me.'"

Along with this understanding, Happy Creek's President knows what the key to everything is. "It really is all about passion," Blann insists. And passion is something this energetic and positive person is never short of. "The thing I'm most focused on is finding a mine. Whatever it takes. I've invested a lot of my own money back into this company and I really believe the properties that we have are certainly some of the best exploration properties going in B.C. today. We're in a good position here; I've had a number of fairly large financing groups express interest in what we're doing. That's great for us. Our moly projects in particular are top notch—and that is our future."

Blann is an adventurous outdoorsman who combined his passions for mountaineering and geology with his interest in mining. "I have a personal motto" Blann laughs, as he travels en route to one of his regular on-site visits, "One foot in the boardroom and one in the bush." Blann is definitely a man who likes to keep his finger firmly on the pulse, in touch with his investments.

HI HO SILVER RESOURCES INC.

President and CEO Fred Fisher has a shrewd business mind lurking behind his friendly facade. Personable, down to earth and amenable, it's no surprise that he charms investors in Canada, as well as those in the U.S.A, Europe and Asia.

A worldly man, Fisher has traveled extensively in his career, but eventually settled back in his home town of Toronto to launch a junior mineral exploration and development company in April 2005. Named 'Hi Ho Silver' in anticipation of owning a large silver property, Fisher admits the company's focus has changed slightly since its inception – now Hi Ho Silver Resources Inc. has set its sights primarily on the acquisition of properties containing molybdenum.

"In the beginning, I was looking for a large silver project mainly, because I like silver - I like the prospects for it and many other aspects to the metal," explains Hi Ho Silver's Fred Fisher, in an interview from his head office in Mississauga, minutes from downtown Toronto. "So that's why I grabbed the name and started around this premise. And then, I was approached with this moly idea and I said, okay, this has a multi-billion dollar potential so let's run with it."

Hi Ho Silver has indeed run with it, and in a short period of time the company has made a name for itself on both sides of the Atlantic and acquired four properties – all presently at the exploration stage. Involving a variety of metals, Hi Ho Silver has found gold, copper, silver, zinc, and lead, in addition to molybdenum deposits, on its properties. "But my main focus is the Carmi property in Kettle River, British Columbia. That's the main one that's probably going to power it for the next little while. It's pretty much purely molybdenum, with a little bit of rhenium," explains Fisher.

"A buddy of mine, Lloyd Brewer who is a really good property guy told me all about moly. I optioned Carmi from St. Elias Mines, at his suggestion. I said, 'Looks like a great project, so I'll take your word for it. I like the size of it.' Even though, at that time, people liked gold a little more; it was Lloyd who pointed out that, in 2005, not a lot of people were paying attention to moly. He said to me, 'As it stands, you can make a case that the property is extremely valuable based on historical reserve estimates based on $25 a pound moly and becomes that much more valuable everytime the price of Moly moves up.

But, this wasn't Fisher's first foray into mining. In a way, mining deals are in Fisher's blood. "My dad John Fisher used to do a lot of mining deals down on Bay Street, in the late 50s, 60s and 70s, so I kind of grew up with them," he says. "I used to meet him downtown at the promoters haunts like Cyrano's or Hy's and listen to all the promoters pitch their deals, it was quite facscinating. I was young but would listen to these characters with nick names like Black Jack, and Kenny "The Wheels" Wheeler, and other assorted handles that make the whole environment quite entertaining. Not quite the same anymore.

"A lot more rules and of course the biggest change was the invention of the internet. "And then when I graduated from university, I just decided I would migrate into that. There wasn't an environment here any more in Toronto, so I went to Vancouver."

The 48-year-old looks back on an interesting and varied career, from handling investor relations and raising venture capital, to holding directorships at several publicly-traded companies.

"I was a stockbroker back in the 1980s, for a few years. From 1986 to 1989, I was at Jefferson Securities in Vancouver, B.C. I caught a couple of years of really good markets, then I caught October 1987; the market crashed and things weren't good for a while after that.

"I remember at the time after the crash, penny stocks and mining stocks in Vancouver were just getting belittled and berated all over the place, and then you came out with your first book – The Wizards: Millionaire Magicians of the Vancouver Stock Market – and it was like a breath of fresh air."

Not long afterwards, in 1990, Fisher returned to his roots in Toronto. "After I moved back, I worked for a couple of companies just doing Investor Relations. But, a friend and I saw the Asian Tiger was growing, so I moved again."

Fisher packed up his bags and headed to Asia, where he lived for over six years, specializing in raising venture capital and investor relations projects. Fisher assisted a number of public companies in developing their business and operating plans and attracting financing for mineral exploration operations by introducing them to potential business partners and venture capitalists.

"I was doing mostly mergers, acquisitions – we were taking OTCBB (Over The Counter Bulletin Board, a securities market in the United States) and NASDAQ deals and merging them with Asian companies. Some of the companies we merged were from mainland China. We also did IR for a few companies there. We just tried to mix and match. It worked out pretty well for us," Fred Fisher says.

Ultimately, Asia agreed with Fred Fisher. "I got used to the hot weather, so now I don't like winters any more. I like traveling, so I do my best to avoid winters – eventually I'm just going to move down to the Caribbean," admits Hi Ho's CEO.

During his six year stay in the East, Fisher became familiar with Asian business, and made many lifelong connections. Yet, a family illness would call Fisher back closer to his native land. "My dad got sick in 2001, so I moved back to Canada to be with my family," he notes.

His father's guidance was partly responsible for his chosen path into moly mining. "When I had graduated with my BA from York University, nothing was really jumping out at me. I worked for the government for a little while – that bored me to tears. My dad said, 'Take your Canadian Securities Course, go out to Vancouver, there's a good market; if some friends are doing a deal, you can participate in that and build it.' That's how I got involved with stocks and mining in B.C. After I got married, we just loaded up the truck and headed out to Vancouver, like the Beverly Hillbillies; heading in the direction of gold and oil, back then."

Vancouver turned out to be the right direction for Fisher – but it was the molybdenum deposits on Canada's West coast, not oil and gold, that would eventually lead to the birth of Hi Ho Silver.

The idea of forming his own company emerged while Fisher was acting president at Golden Hope Resources, a publicly-traded junior gold and base metal mining company, which recently merged with Eternal Energy. "Two friends over at Research Capital in the Vancouver office (an employee-owned, integrated investment dealer based on Bay Street in Toronto), they said to me, 'you should start a new deal.' They said that they thought the CNQ (Canadian Trading and Quotation System stock exchange) was sort of an up-and-coming thing, so – in March of 2005, that's when I decided to start Hi Ho," reveals Fisher.

Born out of his background of leading and assisting public companies, along with an interest in resource exploration and stocks Fred Fisher realized that he had what it takes to make his idea work as a business. "I had the funds from my time over in Asia, and I have good friends who are geologists out in Vancouver that could have access to the properties…" And thus, a new junior moly company was formed.

Now, over two years later, with four properties in Hi Ho's current inventory; Fred Fisher sits back in his chair as he speaks openly about his experiences in the moly and metals business. Their main property is Kettle River, known affectionately as 'Carmi' because of its location in the Okanagan Highlands of south-central B.C., about 45 kilometres south-southeast of Kelowna, near the settlement of Carmi. The company was granted an option to acquire up to a 70% interest in the Carmi property, comprising of 2,873 hectares (more than 11 square miles); they acquired an initial 51% interest.

While Hi Ho Silver has an option to acquire up to 100% interest in each of the three other properties – Tasco, a copper and moly project in central B.C; South Rim, a gold and moly project, also in a mineral-rich region of central B.C; and Silver Tip, lying within the historic Slocan Silver Camp, it's the silver project that Fisher dreamed about.

However, Carmi is the primary for Hi Ho. Kettle River is comprised of 5 mineral claims; located 295 kilometres east of Vancouver, within the Beaverdell Mining Camp – Beaverdell is a mining and logging town; heavy equipment and skilled labour is available locally.

Carmi has good infrastructure, with a gravel road, accessible throughout the property. The terrain is relatively flat and there is potential for open pit mines. Moreover, the moderate climate in the region allows Hi Ho Silver to conduct its exploration work on a year-round basis.

To date, a total of 21,530 metres of drilling has been carried out within the property. As expected, the drilling demonstrated two large areas of molybdenum mineralization occurring within a large-grained crystal, or porphyry, geological environment.

"Carmi is the one that I'll try and spend US$2 to $3 million in the next eight to twelve months on drilling," says Fred Fisher. "Outside of the main one, I can't really say. I'm going to wing it on the other three. I'll likely spend $200-250,000 on each of them this exploration season, and then we'll assess them in the fall and go accordingly."

The moly at Carmi was first discovered in 1960, and companies have been drilling ever since. "The drilling of the ore body, the deposits that we inherited, was done by several companies, but mostly Placer Dome and

Phelps Dodge. They did 188 drill holes; $2 million bucks worth, over the 1980s and early 1990s.

"During their drilling, they delineated two zones that totaled about 23 million tonnes of moly, grading 0.17%. But, they're not 43-101 verifiable, as they're based on historical data. So, I've got to do a certain amount of drilling to reprove them."

Hi Ho Silver's update of the historic resource to the Canadian Securities Administrators' National Instrument 43-101 standards is expected by late 2007. Although, the historic resource estimate (not compliant with NI 43-101) was 20.7 million tonnes at 0.064% molybdenum, according to a 1987 technical report on the property. The data also showed 0.106% molybdenum sulfide or MoS2; the black crystalline sulfide of molybdenum, which occurs as the mineral molybdenite.

But, "what happened was Placer Dome walked away from the project – they walked when moly was $5 a pound and they figured they needed $8 to $10 to make it economical, which is obviously well above," notes Fisher. The Carmi property was uneconomic during the 1990s due to the depressed price of moly, and the mineral claims subsequently lapsed. Then, when moly rose from $2 per pound in 2000, to upwards of $20 per pound, the claims were re-staked in 2001.

However, "They never did any drilling in between the two zones – the two zones are 1800 metres apart. We've spent money in between on a geophysics IP (induced polarization), and we've found two new zones," says Fisher.

"We've drilled two holes in each with all four holes containing both moly and rhenium. So they'll merit a lot more attention and we'll try and link it all up into one big zone."

The discovery of rare and valuable rhenium at Carmi provides an added bonus for the moly explorers. Primarily used in high temperature alloys and catalysts, rhenium is silvery white in colour, and heavy. It was the last naturally-occurring element to be discovered and belongs to the ten most expensive metals on Earth. Total world production is between 40 and 50 tons annually, so this is expected to add meaningful value to the Carmi property.

Previous explorers also reported anomalous values of gold and silver within sections of the zone. There are additionally quartz veins, up to 1.2 metres in width, and sericite and pyrite occurring within the molybdenite and outside of the molybdenite deposit.

Though, Fred Fisher still has his eye on the main moly prize. "I'm hopeful that we have 23 million tonnes of moly, based on old historical data. Our initial objective is to re-confirm the old deposits and then we want to try and link up the two zones, and aim for a hell of a lot more – that's the goal."

Fisher continues, "My geologist is pretty certain that there is a much more serious deposit then the one indicated by the historical data, But, obviously, we've got to do all that work to be able to put that down on a news release... We've done eight holes – totaling 1821 metres – and a geophysics program, and I believe we'll probably end up doing anywhere between 20 and 30 holes. By that time we'll have a pretty good idea of what we've got."

The initial two Hi Ho holes were completed in December 2006, before drilling was temporarily suspended due to severe winter weather conditions. The first drill hole twinned a previously completed 1976 drill hole, and returned 0.107% MoS2 over 173.78 metres, and ended in mineralization.

Furthermore, past work indicates good potential for discovering additional molybdenum mineralization within the Kettle River property, including the likelihood for the discovery of higher-grade mineralization at depth. Hi Ho Silver's own drill program, to date, was successful in confirming grades from previous drilling and confirming the large, previously unknown, new mineralized area discovered by geophysics.

Carmi will be followed up by geological mapping, trenching and diamond drilling during the upcoming field season. Field crews were dispatched to establish additional grid and clear access roads. This will be followed by geological mapping, trenching, further IP surveys and diamond drilling. The work program at Carmi was designed and is supervised by Paul Reynolds, P. Geo. and Michael Sanguinetti, P. Eng.

With Carmi's development and exploration already squarely on his plate, Fred Fisher admits he wasn't actively looking for another property when Tasco was brought to his attention. But, he couldn't pass up such a great opportunity, he says. "It's pretty impressive - grading 0.3% copper, 0.21% moly. A big hole – 1000 ft hole, at the property. It's near the Taseko Lake district, near Taseko Mines (a deposit containing 9.2 million ounces gold and 3.9 billion pounds of copper). And, the Bralorne Pioneer Gold Camp (which has produced in excess 4 million ounces of gold) is only 40 kilometres away. When the guy showed it to me, John Chapman, out in BC (a principal of J.A. Chapman Mining Services); I said 'that's a hard one to resist.' So, I did the deal."

In March 2007, Hi Ho Silver Resources Inc. signed an agreement to earn a 100% interest in the Tasco Copper Moly Project, which is 64 square kilometers, located near Clinton, B.C. The property is also near Galore's mine - Galore Resources Inc recently launched a $1.6 million exploration and drill program on a 48,000 hectare Taseko project.

"I think we're going to do some work together – Hi Ho and Galore. It's also in a seasonal area, where it's difficult to work in the winter. So we're going to share some costs on some geophysics. We're going to reconfirm the Tasco hole and do some more drilling over this summer," says Fisher.

The Tasco property is under explored and an initial $600,000 Phase 1 exploration program, consisting of IP geophysical surveys and diamond drilling, is scheduled to commence this year. To date, eight diamond drilling holes and four percussion drill holes have been completed.

Prior to acquiring Tasco; in February 2007, Hi Ho entered into another property option agreement with St. Elias Mines Ltd., which secured Fisher's company's right to earn a 51% interest in the South Rim Gold / Moly Project, which is located in the Houston area of B.C. "Again, this is St. Elias Mines, connected to my friend Lloyd Brewer."

Comprised of 12 mineral claims, the project covers 53 square kilometres and is located on the south rim of an oval-shaped volcano crater; hence the name 'South Rim.' Fisher reminds that numerous large scale mines/deposits occur in the same region, including Blue Pearl's/Thompson Creek's Davidson molybdenum deposit, which is 145 kilometres to the north, and their Endako mine, which is 150 kilometres to the northeast.

For Hi Ho, the main targets at the South Rim project are molybdenum and gold. The property covers a zone of high-grade molybdenite float grading up to 2.0% MoS2 – found in 1979, by Silver Standard Mines Ltd., within moraine material that was traced for over 2,000 feet south, to the toe of a glacier. As well, there is a gold bearing quartz vein; gold occurs within the vein at 12 separate zones, with values ranging from trace to 24.0 g/t (grams per ton) across 1 metre. Uranium values are also present on the property.

Hi Ho's geologist Paul Reynolds and engineer Michael Sanguinetti recommended to Fisher, based on the presence of significant molybdenum and uranium values in the intrusive rocks and anomalous gold and silver values in epithermal quartz veins; that a two-phase exploration program is warranted for the South Rim project.

This brings us to the last, but by no means least of the four properties; Hi Ho Silver's silver project – 'Silver Tip.' Like South Rim, Silver Tip is blessed with numerous different metals on site; silver, lead and zinc, and some gold. It may be a small property at only 9 square kilometers, but Silver Tip is within the bountiful Slocan Silver Camp, which has produced more than 75 million ounces of silver between 1892 and 1993.

Hi Ho owns 51 % interest in the project, with the option to acquire the remaining 49 %. Located about 50 kilometres north of Nelson, B.C., there is currently a 1.0 metre wide vein, five documented mineralized showings, plus approximately 518 metres of underground development, and high grade silver ore is also present on the property. "There are outcroppings on the mountain; they run like 100 to 1000 ounces per ton silver. Its pretty grassroots, but I'm going to be on there soon too," says Fisher.

"Those three projects are all seasonal. The Carmi one I can work all year round," notes Fisher, of the logistics of drilling and developing these properties. "Carmi is flat, it's just off the highway, it's open pit, and it's cheap. We shut down for January, and the only reason was that this was a particularly brutal winter out there – minus 35. It was getting expensive; drill bits were snapping. Normally we don't have to stop, but the other ones we only explore on a seasonal basis. I think, on the Tasco project, we can probably get in there late May, early June. The other ones are definitely July to early October."

With all of his B.C. properties showing good potential as they sail through development, Fisher admits, "I don't think I'll add any more at this stage – I think there's enough in there to just keep working them." This also leaves ample time for Fisher to apply his business brain to running and growing the company, such as getting it listed on stock exchanges.

Hi Ho Silver went public in August 2006, and their stocks trade over the counter on the CNQ (symbol HIHO). "Ultimately I would like to migrate to the big board Toronto Stock Exchange and am working on a few things to get there. I've got a financing in the works, which will give me enough money. I just need to find a little more continuity in the zones, and a couple of other things," he says.

Things are moving fast for Fisher and his young company. In fact, it hasn't really stopped since he picked up Carmi. "I knew that when I took this moly project on, it came complete with a lot of data and it would develop fairly quickly. I keep kind of winging it, you know, where we go according to how quickly it goes – but it's gone quite a bit quicker than I thought. For a while, I had different companies banging on my door wanting to do different deals. One company from New York, another from Toronto; they're all coming to your door, wanting a position in a moly play before they figure they've missed out.

"You have to be nimble in this business, and then decide quickly as you move along," Fred Fisher says, with certainty. "I raise money but I like to spend it properly – I'm not a big fan of big salaries or anything like that. Make money on your stock and the property will power it.

Fisher concludes, "Just remember, if you don't find the proper property; if you don't develop it into an asset that can be mined; then your stock isn't going anywhere and you're back to where you started."

As a globally-minded international traveler, Fisher knows the world as well as he knows mining. In addition to his CNQ listing, Hi Ho Silver is listed in Germany; trading in Frankfurt under the symbol H9T.

"People, around the globe, are paying attention to molybdenum, as they see it being used all over the place. As I trade quite actively in Frankfurt, I got Hi Ho listed over there; I got listed in November 2006 and, in March 2007, I went to a stock show in Stuttgart and I had a booth there. But, I didn't have to pitch them on moly or anything; they just came over to me," Fred Fisher explains. "I've got a media guy from a public relations company in Germany, but he was reluctant to take it on, because he said many would prefer to invest in gold mining and exploration. But, I said to him, 'Come on, it will fly. Don't worry.' If the size of the project is right, people will pay attention to it. As it goes, German investors took to Hi Ho Silver pretty naturally"

Fred Fisher isn't so much a lone ranger as he is a team player, working with a small staff of professionals. He relies on his Vancouver geologist Paul Reynolds; a graduate of the University of British Columbia, with a B.Sc. degree in Geology. Reynolds has practiced his profession as exploration geologist since graduation in 1987. And, equally so, Fisher turns to his mining engineer Michael Sanguinetti, P. Eng. "He's 70 years old and been around forever. He has worked on most of the major moly projects in B.C. over the years. He's also helped many different companies develop their projects to fruition and is highly respected in the field.

Fisher and two full-time employees are based out of Hi Ho's Mississauga, Ontario headquarters. This includes Isabel Alves, a director since September 2006, working in Investor Relations for the company. She has an extensive background in the financial industry, with years of experience at Edward D. Jones & Co. Ltd. as an operations representative and at Correspondent

Network, a financial consulting services firm in Toronto, as a project coordinator.

As well as Isabel Alves, and President Fred Fisher, the company's five member board of directors includes Kelly Fisher (Fred's sister and company secretary), Thomas Murdoch and Wayne Turgeon.

Thomas Murdoch was the secretary and a director of Golden Hope Resources from 2003 to 2005. From March 1999 to present, he has been the Sales/Program Manager for Matcor Automotive Inc., an automotive parts manufacturer in Brampton, Ontario, specializing in metal stamped and welded assemblies. Mr Murdoch comes from good blue blood mining stock as his grandfather was one of the original founders of Noranda Mining, an old Canadian mining icon. He earned a Bachelors Degree in Economics from Wilfrid Laurier University in Waterloo, Ontario, and a postgraduate Diploma in Business Administration, also from Wilfrid Laurier.

Wayne Turgeon is Hi Ho Silver's Treasurer and Chief Financial Officer. He has been providing accounting, business and tax planning services to individuals and corporations since 1987. With a Bachelor of Commerce degree from Concordia University, Turgeon has previously worked with several large companies, including the Diversey Corp. a subsidiary of Molson, and Deloitte & Touche.

Along with these key personnel, there are occasional field crews. "Sometimes we've got ten people on site, sometimes we have none. I always have my geologists contracted constantly. As well as a mining engineer. Not full-time employees, but part of our team."

As Fred Fisher gets prepared to lead his valuable team into the years ahead, he considers permitting Hi Ho's projects, and other factors. "I don't foresee any Aboriginal claims, especially at Kettle River (Carmi), which is in an old historical mining area. So the permitting, I don't think will be any problem either. Environmental things can come up. You never know until you get further along. At Carmi, the surface is being logged by a logging company; it's being stripped and there are roads going in all the time. If it had been a virgin forest or something, we might have had to go through that as well. But, as far as I know – I will always defer this to my geologist and mining engineer – there should be absolutely no problems. Nothing is in our way right now," he states.

Fisher's long-term game plan for Hi Ho Silver are to take the Carmi property to the pre-feasibility stage as quickly as possible, while the market is hot. "By that time, I don't plan to be the miner though – I'd be looking to peel it off to some bigger guy, or sell it outright to the Chinese, or something."

With his strong business connections in Asia, Fisher is keeping his options open for the future. "I have some decent contacts over there and some ideas, so that might be the way

I might look at. I want to get this fast tracked and then China's my primary place to take this project at the end of it. I think I'd like to pursue the Asian angle."

Already, Hi Ho Silver's recently redesigned website is being updated with translations in Chinese and, of course, German – keeping Hi Ho connected with foreign investors and possibly, one day, prospective buyers of the whole company. "Carmi is the first deal that I've taken from start to finish that I plan to just continue right on through to it's final fruition. I'm not going to try and tell people that I'm going to take these four properties to production ourselves – that's when you hire 200 people, miners, and this and that. I don't want to be there. I just want to swing a deal at some point," says Fisher.

"I've got one of those Blackberry wireless devices – 'Crackberries' they now call them – and I keep it on all night," he says. "I wake up in the middle of the night, when Germany's open and I start responding and checking markets. I have a hard time not paying attention to what I'm doing, until I've figured I've done a proper job."

Dedicated to his business, 24/7, it's difficult to imagine Fred Fisher finding time for any other projects, but he clearly likes to keep himself busy. "I'm starting one other company myself, from scratch, which will go public, the same route of listing on CNQ, Frankfurt, and Toronto hopefully; it's called Chai Cha Na."

Possibly a gold mining and exploration company, though Fisher says, "I haven't really committed yet," he hopes Chai Cha Na Mining Inc. will be trading in the Fall, too. "Then I think the two projects – Chai Cha Na and Hi Ho – will keep me more than busy," he says.

The name 'Chai Cha Na' has an interesting story behind it, just like Hi Ho Silver. "It's actually my nick name in Thailand - it means 'The Man Who Can Do No Wrong,'" he laughs. "A little pompous, I know!"

Nevertheless, with a gold company on the horizon, does Fisher still feel confident in the moly market? "I'm happy with the forecast, definitely! You get kind of swept up in it yourself. I've heard pretty proud predictions for moly, up to $50, maybe $60 a pound. I don't see why it couldn't happen. As far as I'm concerned, if it stays at $30 a pound, that's great! Doesn't really matter, as this project should be pretty valuable between $25 and $30," he concludes.

"With what's going on in Asia, specifically China, and other developments in Dubai - where they're building an entire city from scratch – these constructions will demand a lot of steel and pipelines. Basically, the demand for steel isn't slowing down. "As well there are many new uses for moly being discovered so moly and moly prices should just keep getting stronger for the next however many years."

INTERNATIONAL PBX VENTURES LTD.

Good families stay together, and that's exactly what has kept this Canadian junior mining and exploration company's head above water for so many years. Despite some fairly tough times, President and CEO of International PBX, Gary Medford, is pleased to see that his extended "family" of directors and shareholders has managed to stick around long enough to acquire a molybdenum property worthy of a 'major.'

Now there's no looking back for Vancouver-based IPBX, which has focused on the acquisition and development of moly, copper and gold properties far from home in the South American mining nation of Chile.

"Yes, it's been pretty rough at times, but I think we're now onto something substantial," says Medford, as he prepares for an upcoming trip to Chile, to tour the IPBX properties.

"We've maintained a lot of our original investors—many from Europe. These people have been with us so long. After a while, it's almost like a big family," he admits. "I feel sort of obliged, in the end, to make a pile of money for everybody because they've been supporting and encouraging us for so long. And we also have some very loyal, large American investors too. They've been with us for ten years now. They phone all the time, and we're essentially just one big family here, trying to make a go of something."

Medford's ethos toward shareholders is gratifying and heart warming. As a long-time dedicated investor himself, he realizes that "many companies are set up just to enrich the people that run them. As a result, shareholders can get a rough go." As Medford explains, the goal at IPBX is a little different, " We try to keep everybody happy—not just the insiders. That's why I feel like I have to get this thing to the point where everybody makes a lot of money on it. I think we're almost there, but you never know."

A qualified geologist with a PhD, Medford was born in Montreal and moved out west in 1970. "I worked for Cominco up until 1981. And then I started doing consulting for small mining companies. It was a pretty tough grind as they didn't have much money. Some years, I hardly made anything. So, in 1992, I started investing full-time—that's how I got involved with running IPBX. I was just an investor for many years. I only really took over as president recently."

Medford took the helm of the company in 2006, after years of providing strong financial support. During his years as the 'money man,' he believed in IPBX and delivered a valuable stream of dollars to keep the company in business. Finally, the 59 year old was asked to step up to the plate, as someone who—literally—had so much invested in the junior mining firm; almost as much as the company's founder Verna Wilson, who is the 'heart and soul' of IPBX.

Still sitting on the board of directors at the age of 85, Verna Wilson is the glue that holds the company together. "Verna's here at eight in the morning, or earlier, and works 'til five," says Medford of the happy and outgoing octogenarian. "She's the hardest working of us all, and she kept everybody going through the hard times. She's the optimistic type."

Verna has lived in Vancouver since 1966, but she was born in Ontario. "We moved to Northern Alberta when I was nine years old. So, I'm sort of an Albertan-Ontarian." Verna laughs freely as she talks, showing the good humour that has maintained her spirits, and the spirits of those around her, during the hard times of running an exploration business.

Verna pauses momentarily to adjust the volume on her hearing aid. "I've been a director of companies for the last 35 years. And I started this company 20 years ago. For many of those years, I was the President of IPBX. But then I let a geologist come on as President because I thought, perhaps, it would make more of an impression on shareholders; I'm not sure whether it had an effect or not, but I tried anyway. So, Gary is now President."

Moly is a relatively new commodity for IPBX. "Gold and silver were always looked for, of course. Gradually we evolved into copper, and then we picked up this big moly/copper property," Verna says, referring to 'Copaquire', one of IPBX's four flagship properties in Chile. "The main thrust of promotion of this property has been on the moly because the boys think it's really very large. We won't know until we finish drilling it out."

So far, Verna hasn't yet made it down to where the action is, in Chile; she's too busy holding the fort. "I look after all the administration in Vancouver, so I haven't been out to Chile because I haven't got anybody to take over here. I hope sometime soon."

When Verna originally started the company in 1986, it was known as PBX Resources. "Then, I got involved in 1995," says Medford. "Basically, these little companies, in the old days, it was a lot of promoters running around. It's more sophisticated now. In those days, it was the 'wild west' of investing in mining plays.

"Verna founded the company, and she was very serious about mining and liked the business. But a few people had gotten hold of it who were pretty wild and loose,"Medford continues. "I had speculated in the penny stocks of a lot of different companies; you could make some money. Then in 1995, this company was having a hard time. I bought about $40,000 worth of their stock from some promoter who was pushing it. And I immediately figured, 'Oh jeez, there's another bunch of money down the drain.' But, by making another loan of $25,000, I was told that I could keep the company going."

Gary Medford's investments contributed to keeping PBX Resources alive. But it would be a while before he could confidently put down his cheque book. "When the mining business was starting to heat up again, I thought I better take the bull by the horns because we were getting into some serious mining stuff and it was getting more and more interesting. There was a lot of movement outside of Canada and North America. And I wanted to get something that had more pizzazz than a property in B.C."

To fit its new global focus, PBX became International PBX. "All the other mining companies were spreading out all over the world, looking for new opportunities after the fall of Communism," explains Medford. "One of the opportunities was in Chile. The reason we liked Chile was that it was always rated very highly by the Fraser Institute in Vancouver as a politically stable, good place to do business in mining. Chile was always either number one or number two in the world, alternating with Nevada, and Quebec was maybe third or fourth."

Medford figures, "From a point of view of stability, ease of access to mining properties, and the ability to get things into production, Chile was a good place to go. But, in 1995, Chile was still perceived as 'big company' country—a place for only the majors who were working on really big projects. The assumption was there wouldn't be anything 'small' for small companies like ours to pursue.

"This was especially true in Chile's north. There weren't many juniors there in 1995, and the big mines were projects that only big companies could generally handle; they were mostly up at fairly high altitudes. And a lot of them had been found a long time ago, in the 1970s. In any case, one of the geologists with us had the idea about Region 3, which is somewhat farther south. It was pretty dead in that area, as most of the action was in the north."

According to Medford, Region 3 had been a popular mining area up until the 1920s. "After that, it kind of fell apart and nothing much happened there. But, we could see that there were a lot of opportunities in Region 3. So, Chile seemed like a good bet, and we went for it."

IPBX got their feet wet in Region 3 but they soon boldly entered 'big company' country, and found themselves right at home in Region 1, in northern Chile. They picked up the Copaquire property, which was surrounded on all sides by big company holdings. "It was just by chance that we picked it up. Copaquire was always known as a high grade molybdenum property." Medford believes this bodes well for the junior company's future share prices, as well as increasing the likelihood of a buy-out one day, when the timing's right.

"We've had interest in Copaquire from majors, but the thing is we're well financed right now," he says. "We raised a lot of money because of the interest in moly. We have enough to really develop a resource on Copaquire, to a point where we can increase the value of the property by drilling and expanding the resource rather than trying to make a deal with somebody.

"Basically, we're just at the 'arm waving' stage until we have an engineer actually do the resource for us and we can put it down as a report. So, there's a bit of speculation still at the moment, but right now as far as we're concerned; it's going really, really well."

Certainly, Chile seems to be the promised land for IPBX. The Latin American country, bordered by both the Andes Mountains and the Pacific Ocean, has many distinct advantages over other mining locations, according to Medford. "There are no issues with Aboriginals – that's a big plus," he explains. "Northern Chile is solid mining people." In fact, in most towns in Chile's north, Medford claims, it's not uncommon to find a statue of a miner in the central square. "Everybody is into the game. So, you have a lot of people interested in the business; experienced and wanting to work with you." Except for the company's officers and directors, who are all based in Canada, all of IPBX's employees are Chilean. "Two senior Chilean geologists and three junior ones are working on the drills," says Medford. Moreover, IPBX has a full-time office in Chile, which handles their accounting and project management.

On the environmental front, "If you want to get a mine going, you need to go through an environmental process, but it's dry as a bone out there. There's a limited amount of water flowing, so if you do make a mess; it won't actually go anywhere. Of course, it still takes a year or two to complete the Chilean environmental process, but it's not as tough as North America where everybody's interested in it and they seem to live right exactly where you want to mine something. There's nobody really living in these areas in Chile, so you're not bothering anybody."

Another perk to working in Chile, Medford explains is that "there's a lot of interesting mineralization exposed at the surface—gold, copper, and so on. That's something you'd never find in B.C., for example, without thousands of holes already drilled in it. But in Chile you can find incredibly high-grade veins and structures without a drill hole for fifty miles. It's just incredible exposure, and, of course, it's all desert area; you can see the rocks, there's no vegetation to deal with."

In addition to the copper/moly property of Copaquire, IPBX has three other flagship properties in Chile: Tabaco (copper, gold, silver), Tierra de Oro (gold, copper, silver, cobalt) and Sierra Pintada (copper, gold). Not to mention four more wholly-owned properties in various stages of mapping and sampling to be evaluated for futuredevelopment: Hornitos, Fuego, Palo Negro and Zulema. But, the mania for molybdenum has secured Copaquire's spot as the leading jewel in IPBX's crown of properties.

Moly mania, however, had little to do with IPBX's 2005 acquisition of this renowned 1457-hectare moly property. "Moly mania hadn't really happened at that time," explains Medford. "This property became available—it was owned by one of the wealthy families in Chile, and we got to know them through our lawyer down there. So, we got a really easy deal on it—like I said, it was known as a moly property, but moly was still fairly cheap, back then, so there was no buzz going on about it." At the time of the acquisition, it was Copaquire's copper deposits that Medford and his colleagues were interested in. "We got an opportunity to get in on a property that really a major should own," Medford laughs. "We've got Teck on one side and Rio Tinto on the other. It's a nice chunk of ground, with the potential for a lot of mineralization."

Copaquire is situated in the Andes Mountains of northern Chile, 1450 kilometres north of Santiago and within 20 km of the producing Quebrada Blanca and Collahausi porphyry copper-molybdenum deposits. The property lies in a well mineralized district known to have been worked during the late 1800's. Historic production on the property is estimated to be on the order of 180,000 tonnes, grading about 3.0% copper. In 1973, Placer Dome drilled several holes for molybdenum. According to Medford, "all of them had a huge amount of moly in them, and fairly good copper numbers too." Cominco also drilled the Copaquire property, in 1993. "They got some reasonably good copper numbers by today's standards," says Medford, "about 0.4% copper. That was really what we were interested in when we picked up the property; moly was not of any interest to us, at the time. We already knew they had these fantastic moly holes in there, but it was the Cominco low grade copper holes that made us pick it up." The copper still interests IPBX, of course. The Copaquire property is located near several of the world's largest copper mines, including the Collahuasi mine, which has a mineable reserve of 1.8 billion tonnes at 0.9% Cu.

"Copper is pretty common in Chile and at Copaquire there's a fairly substantial copper deposit, which could be leachable. We drilled at the end of 2006, and we're finding 0.4 to 0.5% grades. Right on a table top mountain; there's 20 metres of dead rock and thenyou get into a layer of enriched copper—so not a pit, but more of a quarry operation. We can see a fairly substantial tonnage developing there. We just need to do more drilling on it," says Medford.

"Meanwhile, the moly started to get exciting so we turned our attention to it—we're trying to get a large resource going, approaching a billion tonnes of reasonably good grade moly." As the drills keep drilling, Medford notes, "It's just a case of punching a lot of holes into it. All of the results we've been getting are very, very good so far. Copaquire could be a molybdenum – copper mine at some point."

The exploration programs to date on the 'Cerro Moly' molybdenum zone at Copaquire have outlined a mineralized area of approximately 900,000 square meters with a vertical range of about 600 metres. At the average specific gravity of 2.6 measured by Placer, that volume translates into a possible 1.4 billion tonnes. The sampling to date suggests that it should be in the range of 0.04—0.06% Mo. The upper 200 metres of this area, which is exposed as a hill, is copper rich, containing up to 0.25% Cu, and somewhat lower inmolybdenum, with grades of about 0.05% Mo.

Another prospective area at Copaquire is 'Sulfato North,' which is part of the same system as Cerro Moly but about 3 kilometres further north and about 200 metres higher topographically. Also part of the copper-rich level of the system, the Sulfato North rocks have been converted from primary copper to secondary (leachable) copper by groundwater action. The material found here is similar in nature to the ore at the nearby Quebrada Blanca Mine, operated by Aur Resources.

The widely spaced Cominco and IPBX drilling at Copaquire has also indicated that a large, secondary sulphide body underlies the Sulfato North area, translating into a possible target of 250 million tonnes of copper mineralization. The copper in this sulphide body is expected to be in the range of 0.4 – 0.45%. The IPBX hole and several Cominco holes appear to have entered the copper-molybdenum shell. This suggests that Sulfato North and Cerro Moly are in fact parts of the same much larger system, with a combined surface expression of about 4 square kilometres.

This exposed porphyry copper-molybdenum alteration system, which exhibits fault related veinlet stockworks and breccias containing copper-molybdenum mineralization, could be as big, tonnage-wise, as the Rosario system at Collahuasi, which contains published resources of 3.1 billion tonnes grading 0.8% Cu and 0.015% Mo. Additionally, just west of the main system, through its recent prospecting and sampling program, IPBX has discovered a second molybdenum bearing system, Marta, which may be at least as big as the known surface expression of the Cerro Moly area.

"We knew that there was potential for good moly and copper," says Medford. "The Copaquire is an option to purchase 100%. We picked up another property, called Tabaco, drilled in 1963 by a small iron ore company. It was reported to have 17 million tonnes of 0.1% copper oxide right at surface, which is interesting—usually you have to go through a bunch of dead rock to get to it. We optioned that too—about a month before copper prices started to rise."

A copper oxide/sulphide-gold-silver skarn property, Tabaco has extensive surface copper oxide showings and workings that extend for up to a kilometre. Consisting of seventeen mining concessions totaling approximately 3,593 hectares, the Tobaco property is located 70 kilometres east of the city of Vallenar and approximately 650 kilometers north of Santiago.

"We drilled enough holes in Tabaco to get a small resource, which amounts to 75 million pounds of copper, officially. We figure there may be about twice as much, using the data from old drill holes." Drilling by IPBX in 2003 reported sections of coarse chalcopyrite up to 76 metres thick and grading 0.85% copper. "And there may be more beneath that," says Medford. "We've got a bit more drilling to do there, but it's one of these properties that's just on the border line of something we can put into production and get a cash flow going."

The Tabaco property covers a 5-kilometre elongate zone of copper, silver and gold-bearing skarned volcanics and sediments. Its extent has been defined by prospecting, geochemistry, trenching and numerous old workings. This zone lies within a major regional fault system which also carries epithermal gold, silver and copper mineralization. In 2003, IPBX drilled holes in Tabaco's Carmen oxide zone, which substantiated the validity of some of the 1962-63 drill results. The drill holes intersected a large copper rich sulphide body with mineralization ranging from 28 metres at 1.69% copper to 144 metres at 0.43% copper.

Like Copaquire, Tabaco has been worked as far back as the 1800s. The property has long been known for its silver and gold, and its copper was high-graded from the 1930s through to about the late 1960s. Ore taken from the Tabaco District, and especially the Tabaco Mine veins, reportedly ran over 25% copper, up to 12,500 g/t silver and from 1.5 to 30 g/t gold. At the moment, according to Medford, "Tabaco is just on the borderline; if we can double the size of the resource—which we think we can—we can make some money on it."

Tierra de Oro is the first property IPBX picked up in Chile. "It was originally a copper property, "Medford explains, "and it covers an area that we thought might have had one of the La Candelaria mines underneath it." La Candelaria was owned by Phelps Dodge and contained over 350 million tonnes of copper ore, grading 0.94%.

But was a copper-rich mine of this kind lurking beneath the surface¿ "We drilled a fair bit, on the copper side of it, until about 1997 when things fell apart in the mining business. We found a lot of copper oxide, but we never did find anything like La Candelaria," says Medford. "But, in the process of staking about 100 square kilometres in that area, we just—by accident—picked up a historic gold mining camp in its entirety. It was sort of sandwiched right in the middle of this area."

Hosting hundreds of small open pit gold workings, the gold camp at Tierra de Oro is reported to have produced several hundred thousand ounces of gold from surface workings prior to the 1920s. It is now considered a strong prospect for the development of a large, bulk tonnage, open-pit type deposit. "As it turned out, according to information we obtained, it was the second largest gold mining camp in Chile, in the early 1900s. It had been mined right up to about 1920 by numerous small miners. The gold mineralization was right at the surface; structurally controlled in a shattered granite body. Back then, small-scale miners weren't availed of the crushing facilities needed to process ore. "So, anything slightly hard they left behind. They almost panned the gold out of the ground—very small scale," says Medford. "We started analyzing some of the samples that were left in the dumps, and we were getting samples up to 260 grams/per tonne of gold – in the stuff that had been thrown away!"

The Tierra de Oro property offers good accessibility, through well maintained gravel roads, and is located only 20 kilometres from the Pan American highway, right next to the Chile's main power grid. With a maximum elevation of only 1,500 metres, the region enjoys suitable climactic conditions for year-round exploration.

Geological, geochemical and geophysical exploration programs to date have evaluated approximately 50 of the 70 square kilometres encompassed by the Tierra de Oro Project. Within this area, many large gold and copper-silver target zones have been discovered.

"We have extensive sampling on Tierra de Oro; with a few holes that were quite good. We never really got a chance to diamond drill it though, because the market fell apart in 1997. We've done a lot of surface sampling since then, and we're ready to drill it again, but, we've gone on to bigger projects that have required our focus because there are big property payments involved. For now, we can afford to keep Tierra de Oro on the backburner until we can get around to doing something with it," explains Medford.

"Essentially, we've redirected our money to Copaquire and Tabaco, because they're under $2 million options to purchase agreements, so we had to spend our funds evaluating them, first and foremost. Now that we know we want to buy those properties, we'll finish the deals and own them 100%."

The last of the four flagship IPBX properties is Sierra Pintada. "Our fourth down, as far as development is concerned," emphasizes Medford. "There are a lot of high grade gold veins and lots of copper on that property." Speaking about the exploration plans at Sierra Pintada Medford says, "We do have targets, but we have to do some more geophysics on that. I think there's a potential for something big underneath. We want to do some induced polarization surveys really soon."

Consisting of fourteen exploration claims totalling approximately 3300 hectares, Sierra Pintada is located in the Chilean Coastal cordilleran copper-gold belt, 650 kilometres north of Santiago. Readily accessible from Vallenar, 30 kilometres to the east, the property covers a 15 kilometer strike length within a several hundred kilometre long major structural zone. Along this structural zone are numerous centres of past high-grade copper, gold, silver and cobalt mining. The Carrizal Alto District, for example, boasts past production of approximately 450,000 tonnes of copper and 1.5 million ounces of gold.

"We also have four lower-grade properties in our inventory," says Medford. "We do a little bit of work on them when we get a chance. But, basically, the lead one now is the moly property, because moly is so hot." Medford speaks animatedly about all the mining projects on his plate, but especially about Copaquire. "It's a very interesting copper property too," he adds. Thus, Medford is assertive that IPBX won't be acquiring any new properties in the near future. "I think for this company, we have enough to make quite a success story—especially with Copaquire. We may take some of our properties and dividend them out into a second company and run two companies instead of just one," he says.

As someone who knows first hand the inherent highs and lows of the mining business, Medford clearly appreciates his change of fortune as IPBX zones in on achieving its goal of rewarding faithful shareholders and keeping the company growing for the long-term.

"We've been really lucky," he says with a smile. "Most of the Chilean properties we've staked ourselves just by being down there. After 1997, everybody disappeared. There was almost no one down there, except us, in that area; as far as juniors are concerned.

"A lot of properties became open ground in the late 1990s. For example, we staked Sierra Pintada just because it became open—nobody had any money to hold on to properties at that time. For a small company, we basically just had enough money to keep the doors open. It was mainly my money keeping it funded.

"And it got worse and worse—there's a minimum cost of about $40,000 a year to keep a public company going. I kept putting more and more money in and the market wouldn't turn. By the time 2002 rolled around, I had a huge amount into IPBX; keeping it alive."

The next step for Medford would have been to mortgage his house, but thankfully, "things turned." He says, "We were waiting and waiting for it to take place. We concentrated on staking things we could afford to stake—because once you have them staked, you just have to pay the taxes on them. When we acquired Tabaco, we were still on the ropes. But, it looked like such an interesting little property, so I managed to drag up another $25,000 to make the first payment on it."

Now, seemingly, the wait is over—with IPBX well established as one of the few juniors in a country of majors. For Medford, and all the others who've devoted money and effort to supporting this junior company for the long haul, including founder Verna Wilson; things have begun to turn.

In the words of Verna, "I'm determined that this one—Copaquire—is going to become a huge success. I just have to stick around until that happens."

KOBEX RESOURCES LTD.

Leo King and Dr. Roman Shklanka, President and Chairman of Kobex Resources respectively, are feeling very lucky about the company's newest investment in one of the planet's largest undeveloped molybdenum deposits.

Situated in Colorado, USA, the magnanimously-titled 'Lucky Jack' moly project—previously known as Mount Emmons—is the center of attention for this Canadian junior resource company. "I feel quite comfortable with stating that I don't know of a better molybdenum property in the world," praises Dr. Shklanka. "It's twice the grade of anything out there that I'm aware of, of that size." Lucky Jack's high grade moly is a boon for Kobex, which has an option to earn up to a 50%, 65% or 100% interest from U.S. Energy Corp. and Crested Corp. Lucky Jack is the culmination of a decade of hard work in the exploration business for the publicly-traded company.

Kobex is helmed by Leo King, who has over 40 years experience in mineral exploration and mining geology, as well as ten years in senior executive positions. In Canada, and internationally, King has managed exploration and development projects and has valuable experience in mineral prospect evaluation and property acquisition. He holds a Masters degree in Geology from the University of Saskatchewan and is a member of the Association of Professional Engineers of Ontario and a member of the Professional Engineers and Geoscientists of British Columbia.

King connected with Dr. Shklanka back in their college days. "Roman and I both grew up in Saskatchewan, and Roman was ahead of me at University. But we first met when he was doing his Ph.D. at Stanford and I was an undergraduate. I worked for Roman, who lead a geological survey crew in northern Saskatchewan, for three summers," says King.

"In my case, I grew up on a small farm in Saskatchewan," adds King. "It wasn't large enough to support my father and his sons so I didn't have much choice—I went to university and, naturally, with outdoor experience on the farm, I chose something close to nature as my profession."

Dr. Roman Shklanka has over 40 years of experience in the mining industry. From 1969 to 1990 he worked for the Placer organization in a variety of positions including General Manager, Exploration for Placer Development, and Vice President–Foreign Exploration for Placer Dome Inc.

Dr. Shklanka was Chairman of Sutton Resources Ltd. and was responsible for the acquisition of Sutton's world-class Bulyanhulu (Tanzania) gold deposit which led to the purchase of Sutton Resources by American Barrick Gold Corp. in 1999 for CAD$525 million.

Shklanka has a doctorate in Petrology and Ore Deposits from Stanford University along with a Master of Science in Geology and a Bachelor of Commerce from the University of Saskatchewan. He was also the Chairman of Canico Resource Corp. until it was sold to CVRD in 2005 for $950 million; he is currently Chairman of International Barytex Resources Ltd. and Polaris Minerals Corp.

King and Shklanka became involved with Kobex Resources and now work together acquiring and exploring properties for the company. "Some people I knew came up to me and said that they wanted to form a company, but they wanted me to help them as they weren't geologists," says Shklanka. "So, they wanted help with target selections. And I said 'sure,' you know, I'd go along with that. We're both experienced geologists; Leo maybe more than me. I've been in administration for so long, I'm not sure that I'm even a geologist anymore," he laughs.

Leo King smiles, shaking his head in disagreement, "I think he is."

"Anyway, things sort of got out of hand, and we went from helping with geology, to doing much more than that," notes Shklanka. "That was about ten years ago."

Today, Kobex hires other people to conduct the field work. "But, the conceptual stuff originates for the most part here," says Shklanka. "We look at a situation and decide whether it's something we want to pursue." And Lucky Jack was one property that Shklanka and King recognized immediately as an opportunity too good to miss.

Located near Crested Butte, Colorado, thirty miles north of Gunnison, Lucky Jack totals 5,465 acres and has the exploration and development potential that Kobex Resources is looking for. The granitic Mount Emmons porphyry intrusion is comprised of several texturally distinct phases of varying relative age. From oldest to youngest, the three principal phases, all porphyritic, are: Red Lady, Keystone, and Union. Red Lady and Keystone are credited with having produced most of the molybdenum mineralization.

The Lucky Jack property has a long history, dating back to the 1960s when the Keystone mine was operating on Mt. Emmons and producing lead, silver and zinc. It was in the 1970s when Amax Inc. began extensive exploration of the site. Amax initially leased the property from U.S. Energy Corp. to determine if a commercial molybdenum resource existed. Eventually, Amax acquired the property outright, with U.S. Energy Corp. retaining a 6% gross royalty.

In 1974, Amax struck moly. Their drilling outlined a large molybdenite deposit—220,166,669 tonnes grading 0.366% MoS2 and outlined at a 0.2% MoS2 cutoff. These numbers, however are based on historical estimates that do not meet today's standards. On March 23, 2007, a National Instrument 43-101 compliant technical report was completed for Kobex by Behre Dolbear, an international minerals industry engineering consulting company. The report has been filed through SEDAR.

"This property is one that's well known by the work that was done in the past, in the late 1970s. In fact, Amax was the premier moly producer at the time when they did the work and took this property to a feasibility study stage," explains Dr. Shklanka. Amax's moly division included the Henderson and Climax moly mines (also in Colorado). In the 1980s, the project was issued permits for a 20,000 tpd (ton per day) block cave mine—a mass mining method that allows for the bulk mining of large, relatively low-grade, ore bodies. Nevertheless, Amax deemed the project uneconomic due to low moly prices and Mt. Emmons was left on the back burner.

Then in the 1990s, there was a new engineering review of the "high-grade section." American Mine Services, Inc. was commissioned to provide a pre-feasibility study. A Behre Dolbear pre-feasibility study was completed for a 6,000 tpd operation based on 22,586,311 tonnes grading 0.701% MoS2 at 0.5% MoS2 cutoff and 10% dilution. It was in 1992 that Amax merged into Cyprus Minerals Company becoming Cyprus Amax. Then, in that same year, Cyprus Amax applied for patented claims on the Mt. Emmons property and commenced securing water rights for commercial production. Delineation of the ore body was determined by 191,000 feet (38 miles) of diamond drilling on the property.

In 1999, Phelps Dodge acquired Cyprus Amax through a hostile takeover. Then in 2002 water rights were granted, and in May 2004 patented claims totaling 365 acres were granted. However, also in 2004, the relationship between the town of Crested Butte and the mining company Phelps Dodge became strained. The locals in the old mining town raised concerns over the safety of their drinking water supplies and the consequences of bringing the "boom-and-bust mining industry" (in the words of the then-mayor) to their resort community. In 2006, Phelps Dodge transferred the property back to its original owner, Wyoming-based U.S. Energy Corp.

Phelps Dodge, along with previous owners, spent over $150 million on Lucky Jack, which is commonly referred to as a 'world class' mineral deposit of molybdenum. To date, the property is pierced by 168 diamond drill holes and its defined resources are described as 'very substantial.' Compare its historical resource estimates with other mines—Henderson has a grade of 0.21% Mo, Climax has a grade of 0.19% Mo, while the Lucky Jack molybdenum deposit reportedly contains 220 million tonnes of mineralization averaging 0.366% MoS2 (0.22% Mo)—950 million pounds of Mo—and high grade sections containing 22 million tonnes at 0.701% MoS2 (0.42% Mo).

Furthermore, considerably significant underground infrastructure is already in place, providing access to the ore body. Roads, power, telephone and water are present at the property which is very attractive to Kobex's shareholders. Existing infrastructure also includes a water treatment plant, commissioned in 1982. As Amax was working to move Mt. Emmons to production, it agreed with the State of Colorado to build a water treatment plant to process mine waters flowing from the old Keystone Mine workings and additional water flows resulting from the ongoing diamond drilling. The plant—with a lime neutralization, flocculation, flotation, filtration system—historically cost one million dollars annually to operate. In 2005, 122.6 million gallons were treated there, and the systems were further upgraded in 2006.

"Logistically, Lucky Jack is very well located, it has great infrastructure, and it's accessed by a paved highway," notes Kobex's Leo King. "There are water resources—there's basically everything that you need. I think we'll probably have to upgrade the power supply eventually."

As per the terms of the option agreement between Kobex and U.S. Energy Corp. and Crested Corp, Kobex will manage development and operations at Lucky Jack. In order to earn an initial 15% interest, $15 million (USD) in expenditures must be spent by Kobex. To earn an additional 35%, a total of $50 million must be spent within a ten-year period from the effective date of the agreement. Once Kobex earns this 50% interest in the property, they will be entitled to enter into a 50-50 joint venture with U.S. Energy Corp. and Crested Corp., or possibly take over the project 100%.

"Our first letter of agreement was signed in October of 2006. But, it was subject to a 60- day due diligence. We then worked out the final agreement and that was signed in the Spring of 2007," notes Shklanka.

King adds, "We plan to issue a new resource reserve estimate, once we've had a chance to carry out an extensive underground drilling program. But, the historic resources obviously suggest a large resource is there. Within that larger tonnage—which on its own has a respectable grade, compared to other producers and other deposits—there's a higher grade resource that they identified of about 22.5 million tonnes grading 0.7% MoS2. And that, again, stands out as very high grade compared to other moly properties that we know of."

Kobex has a timeline in mind for the next steps to be taken at Lucky Jack, including initiating environmental work and permitting, establishing effective relations with the local community, and a full feasibility study. "Initially, we need to rehabilitate access to the underground drilling stations that Amax used," says King. "Rehabilitation is necessary because that work was done about 25 years ago, so we have to make sure that there's safe access to the drilling stations.

"To do this, we have to carry out an underground rehabilitation program. This will be our initial step. That will probably take us to the end of the year, 2007. Following that, early in 2008, we hope to be starting a major underground drilling program to better define the resource, focusing on the higher grade portions first and foremost," concludes King. The company expects additional resource delineation with future drilling.

Looking even further ahead, by 2010 Kobex plans to have a new feasibility study completed and permits issued. Also a priority by 2010 is completion of an Environmental Impact Statement (EIS), including a socioeconomic study. Their ultimate objective is to begin mine and mill construction in the year 2011 and to be a producing moly mine by 2013.

"We have all the money we need to take this project to feasibility, and through feasibility," notes Dr. Shklanka.

Meanwhile, King adds, "From an environmental point of view, we believe that this can be permitted and, in fact, historically Amax did permit this property for a 20,000-ton-a- day operation, though due to low metal prices at the time they did not sign the final permit. But, in the past, this property has been permitted and we expect that, if we carefully follow the permitting procedure and make sure that we engage the public during this process, we will be able to permit Lucky Jack."

To help ensure the permitting and environmental work go smoothly, Kobex relies on its Manager of Environmental and Regulatory Affairs, Clyde Gillespie. Gillespie was formerly a Kinross Gold Project Manager

(2004-2006), and, also with Kinross, he was Environmental Manager of the Ft. Knox Gold Mine, near Fairbanks, Alaska (1999-2004). Additionally, Gillespie was responsible for permitting Buckhorn Gold Mine (North Central Washington State) in 2006, and as Project Manager of the Twin Creeks open pit mine for Santa Fe Pacific Gold from 1994-1997, he successfully permitted the mine expansion. He also assisted drafting an EIS for the Round Mountain Gold Corporation expansion (Nevada 1993-1994).

Moreover, Patrick Moore—a co-founder and former leader of Greenpeace—and Tom Tevlin, who are the Chairman and President of Greenspirit Strategies, have been retained as advisors to assist in keeping a balanced approach to public and community relations for the Lucky Jack project. "We've engaged Patrick Moore and Greenspirit Strategies for a number of reasons, but one is that they've got an inside track when it comes to what the NGO's might be doing and we're certainly conscious that they could get on our case at any time," says Dr. Roman Shklanka. "Someone like Patrick would be in a position, a better position than we are, to have a dialogue with them. So Greenspirit Strategies will be providing an overview of our environmental program."

Furthermore, Kobex's experienced management team has a proven track record of value enhancement and they have played instrumental roles in identifying mineral resource opportunities that have resulted in numerous producing mines worldwide. "There's Dan Munter who is the consulting project geologist at the site," says King. "He will be advising on directing the drilling program and the compilation of the geological model at Lucky Jack. He's a very experienced geologist and he's worked for a number of large companies, including BHP. Dan has considerable experience in copper-moly systems. So, he's a key member of the team with a lot of experience. We have just hired Michael McInnis, a native of Colorado, as chief geologist. He has considerable experience in managing drilling programs.

"Jim O'Rourke is key as well; he's an experienced mining engineer, who was involved in a number of major mine start-ups," says King. A graduate from the University of British Columbia with a BSc. in Mining Engineering, O'Rourke was involved in a number of major new mine start-ups during the first 14 years of his career, including Gilbraltar, Marcopper, and Endako. As former President and CEO of Princeton Mining Corporation, O'Rourke was responsible for the acquisition of the 25,000-tpd Similco open pit copper mine in British Columbia. He was also involved in the development of the $60 million underground block cave development at Cassiar and the $140 million Huckleberry mine.

"Jim is a very well known, dynamic individual and because of his molybdenum mining and processing experience; he's a valuable board member," King advises.

"The other directors have, I guess, a lesser role in the company, although there's Steve Bebek, a business man, who was one of the original founders, along with Evan Koblanski, the other founder of Kobex."

Steve Bebek is a self-employed developer with over 25 years experience in real estate development in West Vancouver. For 15 of those past years, Bebek has been involved in the financial markets and has extensive experience in raising funds for junior companies. His co-founder, Evan Koblanski, is also still on the board of directors. Koblanski brings many years of corporate management experience involving a diverse variety of activities and industries.

During the 1970s, Koblanski was President, Director and part owner of Kolstar Development Ltd., a private company that owned and operated numerous hotels and buildings in British Columbia and Alberta. Since that time, he has been involved in several mining related projects in Mexico and elsewhere in Central America. He was President and Director of Lytton Minerals Ltd., which explored a major open-pit copper deposit in Mexico. Lytton Minerals later became an affiliate, first of Patino Mining Corporation and subsequently of Hudson Bay Mining and Smelting Ltd., itself a subsidiary of Anglo-America Corp.

"Then there is Geoff Bird, the corporate lawyer and a Director of the company," notes King. Bird has been practicing in the areas of corporate, mining and securities law since 1972. Prior to becoming a lawyer, Bird earned a BSc. degree (Honors, Geology) from the University of British Columbia in 1965 and worked for three years as an exploration and mine geologist.

"Also, we've been gearing up to start or initiate this feasibility study at Lucky Jack. We've recently hired Maurice Tagami, who's a metallurgical process engineer by training. He's our Chief Operating Officer and he will be directing the feasibility study work," says King. He continues, "Roman has employed Maurice on two other ventures, the most recent one was with Canico on a nickel project in Brazil. And so Maurice is very experienced, having been a senior project manager for the Onca Puma nickel project. We are very lucky to have him join our team." Tagami has over 26 years experience in mine development and operations; he specializes in extractive metallurgy, and mineral processing project and development management at existing mining operations, as well as implementation of process facilities for new base metal projects.

"In addition, we've recently appointed Andrea Zaradic. She was also involved in the Onca Puma project and is a very capable engineer. Those are the two key employees that we've hired for our upcoming development," says King.

Of course, with such a skilled team behind them; Lucky Jack isn't Kobex Resources' only foray into mining exploration. Previously, the company acquired options to explore various other properties. The Morley River Property in the Watson Lake mining district of the Yukon territory; the Titan Property in B.C.; the Cumo Molybdenum Property in Boise County, Idaho; and the Trikay Copper-Gold Property in Queensland, Australia; to name a few.

In 2003, Kobex entered into an agreement to acquire a 50% interest in the Morley River Base-Metals Gold project. The immediate target on the three claim groups was a VMS deposit (volcanogenic massive sulphide). In 2004 the company entered into another option agreement for the B.C.-based Titan molybdenum property, to acquire up to a 60% interest. Yet, in 2005, Kobex decided to terminate both option agreements.

"So, no, this isn't our first moly project," explains Roman Shklanka of Kobex. "We were also involved, but not for very long, in one in Idaho. We dropped it when Lucky Jack came along."

The Cumo, and the Trikay, were part of a 2005 option agreement with Mosquito Consolidated Gold Mines Limited for Kobex to acquire a 100% interest in two Mosquito properties. The Cumo was a moly project and the Trikay was copper-gold, located in Queensland, Australia, for which an exploration permit application had been filed. But, by 2006, Kobex had abandoned its interest in these properties.

"The Trikay in Australia; we gave that up fairly quickly," says Shklanka. "And it basically came down to the Cumo property in Idaho; when the Lucky Jack came along, the requirements that we had to meet for the Cumo property weren't justified for working them both. So we're currently just involved in the one," says Shklanka.

However, judging by this role-call of international locations, it is apparent that Kobex has taken a global perspective to mining opportunities, from their current Colorado property to finding prospects in several other countries. "I wouldn't say that was the mining strategy initially," notes Shklanka. "It's just the way it worked out. I believe that we were open to anything, anywhere, providing we like what we see. I think the odds against you, in this business, are pretty high and if you start restricting yourself you just increase those odds."

Aside from Kobex's involvement with these preceding properties, in 2004 the Canadian company also entered into a cooperative agreement with another company—Pacific Imperial Mines Inc. —to work on a resource property in China and form a joint venture. In order to acquire an interest, Kobex would have been required to finance $3 million of exploration and development expenditures within three years of identifying a suitable property for this purpose.

"We had an arrangement, whereby we expected to get exposure to China," says Shklanka. "Pacific Imperial would have more than they could handle, so we looked at having Kobex have the right to get something that they may come up with. But, tragically, the main push behind the company, from China, unfortunately was killed in a car accident shortly after we had made the arrangement. That, and other factors, played into us not having exposure to anything that we thought we wanted to get involved in." This agreement expired in August of 2007.

Nevertheless, the Kobex President and Chairman declare that they haven't necessarily ruled out joint ventures in the future, as a possibility, though Shklanka admits, "it depends who you talk to in the company."

"Joint ventures, generally, represent particular challenges," contends King. "Let's say—different personalities, different philosophies. Some of them work well, some of them don't. But, ideally, I think we would be happier if we could acquire properties that are not a joint venture. Because it makes life a little simpler," he adds with a smile.

Acquisitions, though, are presently far from the minds of Kobex's management. Dr. Shklanka asserts, "I would be hesitant to dilute the company just to have more properties in the pipeline, because I think we have something that is very significant. Of course, if something better comes along, we'll look at that very seriously. But, right now, I believe we have something we can get our teeth into and we can concentrate on."

Unquestionably, the exploration game is rarely guaranteed; often fraught with ups and downs and uncertainties. Thus the men behind Kobex are hopeful for good prospects from Lucky Jack—which has the strong potential to be the lowest cost producer for a primary molybdenum deposit of its size.

"Definitely, sometimes there's disappointment early on," asserts King. "You acquire a property and you have an initial look, or carry out some initial work and it doesn't prove up to what you expected. Then, you have to make a decision to drop those, and that was the case with the initial moly property that we had in northwestern B.C., and then also with the Cumo."

"What hasn't been mentioned about Lucky Jack is that this is a pure moly play, as opposed to a by-product moly," says Dr. Shklanka. "The advantage of a pure moly play like this is that you generally get a first-rate product. It doesn't come with all the 'nasties.' As such, it is a preferred product for roasting and hopefully could command some premium in the marketplace."

Shklanka elaborates, "In the past, moly was very vulnerable price-wise, and I think that was primarily due to the fact that most of the moly coming into the marketplace was—to a large extent—by-product moly. Mines were producing it, regardless of how much of it was coming out, just because they were producing copper and making their money on the copper. The price of moly didn't affect whether it was going to be put on the market or not.

"Yet, I think we've gone beyond that now—where you need additional production coming on stream, which isn't by-product production, in order to meet the requirements in the future. And in that respect, I think we're well positioned to be there," he concludes.

Well positioned indeed. "I don't know of a better moly property," reiterates Shklanka. "Lucky Jack comes in at twice the grade of competing properties. As a result, I believe it will be a low, if not one of the lowest cost producers out there. So, I have a lot of confidence that this is going to succeed in the end."

As Kobex moves forward with their promising Lucky Jack moly project, the predictions for moly market prices seem poised to add further value to the high grade moly deposit's expected worth. "The future for moly—well, you know, I'm a geologist, not a metals predictor," says Shklanka, "but, the indication is that the steel industry is going to be in continued demand for quite some time. The expectations of China and India and some of the other developing countries, as well as pipelines. What I've been exposed to, for the most part, would suggest the continuing demand to remain strong.

"One hard number I've seen advises, by so-called experts, of 4.2% growth. A 4 to 5% growth rate is quite significant. That's certainly a lot higher than the normal world growth rate, so that would indicate to me that a deposit of this type would have to come on stream every year, in order to just keep up with the increased demand that is out there," Shklanka notes. "And I have a hard time seeing that happening, so I think that moly—over the medium term at any rate—should have a good future."

Leo King leans forward to add his viewpoint. "We're pretty positive when looking at the future of moly. And as Roman said, it has some very key uses, particularly in the steel industry and very specifically in specialty steels. Again, you know with the pipelines having to be replaced, certainly in North America, some of these major projects—if they come on stream—there would be a huge demand for moly-alloy steels."

The likelihood of a bright outlook for moly combined with a large moly property called Lucky Jack—this all amounts to confidence being high at Kobex that things will pan out this time around. Add to this mix the capable and cohesive President Leo King and Chairman Dr. Roman Shklanka, with their geological expertise and years of mining experience, and the stars seem to be aligning for this fateful property.

MOLYCOR GOLD CORP

President Ed Lee admits that the company's name, 'Molycor Gold,' might be misleading. "Yes, we're about moly and gold; a little bit of both. But, it is a misnomer," says Lee. Essentially, the Vancouver-based junior exploration and development company dabbles in a variety of metals including molybdenum, gold, silver, platinum, palladium and copper.

Molycor has acquired a plethora of properties, principally in British Columbia, Canada and Nevada, USA. "We're at the exploration stage with Crowrea, in BC, and the Nevada packages," explains Lee, "In fact, the Nevada ones have taken a while to put together. We've been actually putting that package together for over two years. And we're at the point now where we've completed it, and have several drill targets. So, this year alone, we could end up with two, three, four—if not five—different drill programs happening, on three of our properties."

Crowrea is Molycor's chief molybdenum property; one that is stirring up excitement among directors and shareholders. Yet, Edward Lee isn't ruffled by the ensuing 'moly mania.' Instead he remains cool, calm and collected. "I'm an easy going guy," he says. "I'm an oddity, apparently. It just seems to be that my outlook is a little different from most."

Moreover, reflecting on the company's polymetallic portfolio, Lee has a business mantra, "Whatever is going to make the company money."

Ed Lee became President of Molycor in May 2005. He is also Director and Executive Vice President for Adanac Moly Corp. and a Director of both Goldrea Resources Corp. and Rocher DeBoule Minerals Corp.,—all owned by Larry Reaugh. "Larry has put together four companies," notes Lee. "They are four individual companies with their own individual projects." Reaugh has forty-one years experience in the mining industry with the past 25 years directly involved in mineral exploration with junior resource companies.

Now Lee presides over Molycor, while the former-President, Reaugh, remains Chief Executive Officer—since 1981—and Chairman. "I'm Larry's right-hand man. I've been working with Larry since 2003. Larry and I hit it off when we first met. We saw eye to eye. I've been a Director of Molycor since July, 2003. Also, I'm the only person that Larry has had on—other than Teresa Piorun, the Corporate Secretary—as a Director on all of his companies, in the 30 years that he's been doing this," adds Lee. Lee also has ten years experience in the assistance of financing and corporate development of public companies, as an entrepreneur in private business in northern British Columbia.

"I was born in Vancouver, raised in Kamloops, and lived in Fort St. John for 20 years. But, now I'm back," he says. Undoubtedly, Lee couldn't be happier than he is in White Rock, the British Columbia base for Molycor's office. "The office overlooks the ocean, there's a big beach. And there's a boardwalk that's absolutely gorgeous. It's wonderful!" he laughs. "It could be pouring rain in Vancouver and be sunny here." This reminds Lee of another mantra, equally appropriate in life and the mining industry; "Timing is everything," Lee says.

And the time certainly seems to be right for Molycor. "Molycor has been around for a long time—over a quarter of a century. It was around long before me," says the 45-year-old President. But in his two years as President, Ed Lee has seen the company's steady, continuing advancement of several key properties. "When I first got involved here, it was over a project called the Windpass, which we still have. But, we needed something more. So, you know, we've been working on that, and putting things together."

Currently, Molycor Gold Corp. has many active exploration projects in North America, including B.C.'s Dobbin I (gold, copper, platinum and palladium), Dobbin II (known as the Tadpole Lake molybdenum project), the Crowrea moly property, along with properties in the Battle Mountain-Eureka gold trend in Nevada.

The Crowrea, in particular, receives a concentration of Molycor's time and effort. "Yes, the Crowrea's got a pretty good upside to it, in that it could be—at this stage—a small production," says Lee. "Or, eventually, it could be a large production. We just don't know; we need to do quite a bit more drilling. But, what we're looking at is, basically, there's a good chance that we could be doing anywhere between 200 to 1,500 tonnes a day, let's say, for the next five years or more.

"So, you know, the way it's kind of shaping up is, there's the potential there for a dyke that we've identified," says Lee. Crowrea is a mineral rich occurrence, which was interpreted as a northeast trending dyke zone. "It could be extending over 1,000 metres," suggests Lee. "Well, if that's the case, then we have something here that could be around for the next ten years, generating some good income for the company. When money is coming in, that opens the doors for doing other things. So you're not having to do financings all the time."

Crowrea was first discovered by Reaugh's Goldrea company (then known as Verdstone Gold Corporation) in 1995. The Crowrea prospect, approximately 3,440 hectares (8,500 acres) in size, is located near Summerland, B.C., 15 miles south of Brenda Mines, which produced copper and molybdenum for 20 years prior to closing its doors. Crowrea is currently a 50/50 joint venture partnership between Goldrea and Molycor Gold Corp. and was acquired in 1995. The discovery of mineralization at Crowrea was contained in a trench which returned 0.125% Mo (0.209% MoS2) over 32 metres. Subsequent drilling returned up to 0.164% Mo (0.273% MoS2) over 39 metres in a hole numbered '95-03-09' and 4.6 metres grading 2.034% Mo (3.390% MoS2) in hole '96-14-56.'

During the summer of 2005, the companies completed a close-spaced diamond drill program, with drilling based on a grid pattern with 10-metre centres. The goal was to identify a localized high grade lens of molybdenum, and to test for the possibility of a 10,000-ton bulk sample. This drill program identified the northwest trending dyke; the initial 12 holes intersected favourable molybdenum bearing phases within older, coarse-grained quartz monzonite host rock. Ultimately, the trend potentially extends 450 metres southeast by 650 metres northwest, giving an overall strike length of approximately 1100 metres (3600 feet). The program was managed by Dick Addison, P.Eng., a qualified person under NI 43-101.

Meanwhile, in April 2007, the joint venture of Molycor and Goldrea announced the acquisition of the 'HP' molybdenum claims, consisting of 82 claim units totaling 1629.8 hectares (4027.5 acres). Advantageously, the HP area adjoins the northwest border of the Crowrea.

Cominco Inc. originally acquired the HP claims in 1979 and performed soil and geochemical surveys analyzing for copper, molybdenum, zinc, and lead, resulting in the identification of several molybdenum soil anomalies. Consequently, in 1980 Cominco drilled two percussion holes to a depth of 195 metres, encountering a maximum grade of 0.091% Mo near the bottom of hole 'PDH-HP-80-1.'

This recent acquisition of the HP claims by Molycor and Goldrea isn't the first—Molycor and Verdstone (Goldrea's precursor) jointly acquired the property back in 1995, at the same time that Crowrea was picked up. In 1997 they carried out geochemical and geophysical surveys, adding more molybdenum soil anomalies to the ones previously identified by Cominco.

Yet, the property was subsequently dropped by Molycor and Verdstone due to the low price of moly.

However, it is believed that by reacquiring the HP claims, Molycor will increase the total number of potential molybdenum occurrences—with the addition of HP's 'Webb Site,' 'Noranda Showing,' 'Swamp Anomaly' and 'Empress'—within the Crowrea moly camp. This acquisition brings Crowrea's total to 284 units, amassing 5833 hectares (14,414 acres) of ground.

Planned for the summer of 2007, the companies have outlined diamond drilling programs amounting up to $1 million in work to test the Crowrea and the Empress molybdenum acquisition. Phase I drilling on the Crowrea is designed to infill along the potential strike of the dyke, which stretches from the Noranda Showing to the Swamp Anomaly. Moreover, the drilling program includes holes to be drilled on strike into the Swamp Anomaly, which is approximately 600 metres in diameter, and was initially identified by a geochemical survey carried out in 1995.

Situated approximately eight kilometres within the Crowrea claim unit, the Empress property covers a 730- by 360-metre area, consisting of widespread, low-grade molybdenum and copper-bearing sulphide mineralization with higher grade anomalies. The Empress program will consist of trenching and mapping the attitudes of the veins and fractures followed by diamond drilling. The Empress is classified as a "Mo-Cu-Au porphyry characterized by disseminated, fracture fill, and quartz vein sulphide mineralization."

Historically, from 1968 to 1970 Anaconda Canada Explorations Ltd. carried out work on the Crowrea site consisting of geological mapping; soil and silt sampling; IP, ground magnetometer, and refraction seismic surveys; trenching; and diamond drilling. This was followed, in 1981, by an exploration program consisting of 3428 meters of percussion drilling in 42 holes which yielded grades up to 0.05% Mo over 48.0 metres. But, it is thought that the potential "running off" of molybdenum may have occurred during percussion drilling, and therefore Molycor and Goldrea believe that grades will improve with the use of a diamond drill.

While Molycor focuses on the molybdenum values at Crowrea and its neighbouring HP claims, the company's additional B.C.-based properties—Beaverdell, Dobbin I, Dobbin II (another moly prospect), Flap, and Windpass/Sweethome—are being placed on standby.

"They stay on the backburner until the right situation comes along," says Molycor President, Lee. "The biggest thing with all our properties is we don't owe anything on them. We don't have option payments, or agreements, or arrangements. We own them all 100%, with the exception of the Goldrea joint ventures. So, the only thing we pay is when we do a little work on them, which can be minimal, or we pay the land fees on them."

To summarize, the 100% Molycor-owned Beaverdell claims are located three kilometres south-southeast of Beaverdell, B.C. and 0.5 kilometres south of Teck Cominco's past producing Beaverdell mine. With silver rich veins found in a 3.0- by 0.8-kilometre belt, the claims are referred to as the Beaverdell silver-lead-zinc vein camp. The Beaverdell mine actually produced almost continuously from 1913 to 1991, mining 1,321,140 tonnes and producing 34,594,385 ounces of silver, 16,725 ounces of gold, 25,569,475 pounds of lead, 30,644,112 pounds of zinc, 128,244 pounds of cadmium and 25,699 pounds of copper, according to the MINFILE Production Report. (MINFILE contains geological location and economic information on over 12,000 metallic, industrial mineral and coal mines, deposits and occurrences in B.C.).

Molycor also acquired more claims to the north of the Beaverdell Silver Camp, increasing the size of the overall property by 15% to include key ground of Teck Cominco's silver-producing Highland-Bell Mine and vein system. It is believed that this has improved the property's lead, zinc and silver potential. Chip samples from 2004 showed grades ranging from 192 to 4805 grams of silver per ton.

The Dobbin I project is a polymetallic platinum, palladium and copper prospect, located 27 kilometres northwest of Kelowna, B.C., and adjacent to the Dobbin II 'Tadpole Lake' property. The Dobbin I is a 50/50 joint venture with Goldrea, as is the Dobbin II—a molybdenum prospect. Formerly, Cominco held the Dobbin II ground for 20 years. Between 1977 and 1980 they completed a total of 73 percussion holes (19,026 feet) identifying a significantly sized anomaly. The property hosts a quartz stock work and molybdenum soils anomaly measuring approximately 1,000 by 1,500 metres. The Cominco drilling, in 1978, included widely spaced percussion drill holes with results up to 0.105% $MoS2$ over 56 metres, making this property an important focus for the future.

In addition to the two Dobbins, Molycor co-owns (with Goldrea) the Flap Project, a gold prospect in the Nicola Mining Division of southern BC, and owns outright the Windpass and Sweethome properties, also gold properties, in the Thompson Plateau area of central B.C. The Windpass and Sweethome are past gold producers with Windpass producing 34,456 ounces of gold from 102,996 tonnes, milled over a six year period from 1933 to 1939. Recent drilling on the Windpass property identified a vein structure with assays of 1.5 metres at 0.58 grams Au and 1.6 metres at 1.34 grams Au. Molycor has also staked an additional 18 units to cover the 'Gassen' zone to the west and north of the existing leases.

In British Columbia, there is often the likelihood of Aboriginal claims. "You have to be upfront and honest with everybody. Same as your shareholders, you've got to be upfront with the stakeholders, whether they are Aboriginal or not. And your approach has to be very forthright. You start doing the underhanded stuff, there's just no room for it anymore; you're going to have all sorts of serious issues. The key is to be completely honest with everybody and let people know what's going on. Keep them informed," says Lee. "That's the way I run my business, you know. If you have a problem, I don't know if you don't tell me. So, you've got to be able to reciprocate that—they won't know if you don't tell them. Therefore, keeping people informed is really the key."

With the majority of Molycor's B.C. properties on the backburner for now, though, Lee and his team are able to pay special attention to their acquisitions south of the border, in Nevada.

Holding a total of 462 claims, totaling 9240 acres, Molycor has six separate exploration properties in Nevada, each with gold or silver and base metal potential. They are named Griffon, Ridge Top, Jo, TKO, Hot Dog Ridge, Silverado and the Tami-Mosi claims and all are 100% owned by Molycor. The properties have shown similar geological features and are comparable to the famous Battle Mountain-Eureka gold trend, located 60 kilometres to the northwest.

The Griffon property is 46 miles southwest of Ely, Nevada, and is made up of 82 unpatented claims in the southern portion of the Battle Mountain gold trend. Billiton Minerals originally discovered gold mineralization at Griffon in 1988 at the 'Discovery Ridge' area. Billiton and Westmont Gold established a reserve of 59,459 ounces of gold from 1.71 million tonnes of material. The claims were then transferred to Alta Gold in 1994, who worked the Hammer Ridge deposit with a reserve of 32,907 ounces of gold from 1.32 million tonnes of material. Alta Gold took the Griffon into production in 1998 and increased reserves to approximately 100,000 ounces. The projected mine life was one year and up, producing 1,000 ounces per week at a cash cost of US$145 per ounce. Alta Gold subsequently produced

90,000 ounces of gold from shallow, low-grade oxidized ore in the Griffon's Chainman Shale. Now Molycor has established, based on a review of available data and discussions with geologists familiar with the project, that excellent exploration targets remain at the under-explored Griffon site. And field work followed by drill permit applications will take place throughout 2007.

The Ridge Top, TKO, and Hot Dog Ridge mineral properties, which consist of 30 lode claims covering approximately 2.9 kilometres, are also of great interest. Situated 30 kilometres southeast of Ely, in White Pine County, the TKO claims have particular gold potential in the form of two drill holes by Alta Gold in 1993, located 600 metres apart. It is noteworthy that AmSelco, a former U.S. subsidiary of BP Selco, held the Ridge Top claims in good standing for two decades, and there are indications that considerable development had occurred. Thus, Molycor is endeavouring to acquire the resulting trenching and drilling data.

Molycor's Silverado project is based in the Pinto Mining district consisting of 169 claims, at the southern end of the gold-rich Battle Mountain-Eureka trend, with similar geological structures as Barrick's West Archimedes/ Ruby Hill mine and is located 12 miles to the northwest, which operated until 2002. The Ruby Hill mine is set to restart on the East Archimedes deposit, which has reserves of 1.0 million ounces of gold. Barrick is expected to start production in 2007.

While at the Tami-Mosi property— a past producer of gold, silver and manganese located eight miles southeast of Ely, Nevada, in the Tamberlaine district—work has identified northeast and northwest trending mineralized zones. And, in late 2006, a biogeochemical survey was carried out to test for anomalies in the sediment overlying the intersecting mineralized trends. Ongoing work is planned to continue testing this structural intersection as well as to test geologically for along-strike and down-dip extensions of surface mineralization. Molycor recently staked an additional 78 unpatented claims at the Tami-Mosi property bringing the total number of Molycor controlled claims in the district to 119. All new claims were acquired for the cost of staking and are subject to a 2% Net Smelter Return.

Detailed sampling and mapping is continuing on all of Molycor's Nevada properties to identify gold and silver bearing structures. "We've got a field geologist down there, doing all our Nevada work," says Lee. "As well, we formed a U.S. subsidiary for Molycor."

With a small Nevada office of two employees— "a secretary and one man runs our subsidiary company" —the rest of the business is handled in

Canada at White Rock, where Molycor, along with the rest of Larry Reaugh's group of companies, share a core staff, under one roof. "We've got some accountants here that share duties, and we have our corporate secretaries, our legal secretaries and a couple of others also," explains Lee. "We've just hired an in-house geologist. He's working on all our maps and he's going to be quite busy. He's working pretty much for Molycor. We also have a couple of independent geologists that work for us."

Supported by Molycor's B.C. headquarters and its small state-side subsidiary, Lee admits that he doesn't have any qualms about handling the Nevada properties, despite America's reputation for being stricter on permitting. "Actually, some people will complain about the permitting processes in the U.S.—Nevada in particular. But, the one thing that is kind of nice is the fact that you're dealing with a country that has 300 million people, and the rules are well laid out. So, you know what the rules are before you get started. You can plan your permitting processes along with your drilling plans, your programs. You're not guessing. And there doesn't seem to be a lot of subjectivity. It's pretty straightforward."

Although, he notes, "There is one permit that takes a little longer, because it deals with U.S. Federal Forestry land. That includes the U.S. Environmental Protection Agency, and so it is a little longer process. But, like I said, as long as one knows the rules of the game and the time frame it takes, then you just schedule around it. So, we're going to be looking to have those permits by the end of the fall of 2007. And we're looking to start drilling on the Tami-Mosi by the fall. We've already got drillers and everybody else lined up to go. Thus, when one ends, the next one starts." Unlike British Columbia, in the desert of Nevada, peak season almost never ends as it is unimpeded by Canadian winter weather. "Yeah, we can drill pretty much year round on the Nevada properties."

Regardless of temperatures, Molycor is steadily developing the mining potential of its interests on either side of the U.S./Canadian border. President Ed Lee says, "It would be nice, between the moly property and the gold properties, to actually start closing in...have some kind of a resource put together. And, that way we can move forward."

This tandem approach of exploring both molybdenum and gold properties goes hand-in-hand toward achieving Molycor's ultimate aims for prolonged and stable success. "I would agree that the main thrust of the company is the moly, but the reality is we've not only got a pretty good moly situation, we've got a pretty good gold situation as well. Molycor definitely has support from both sides," explains Lee. "Our biggest asset is flexibility and diversity. The one thing about Molycor is that you're dealing with two metals to work off of. Good metals, too. I mean, they're both strong, and—like anything in the exploration game—if you punch a good hole, it makes all the difference in the world. So, when we punch some holes and get any half-decent results, this thing's going to look very promising."

He continues, "I come from a private business background, so one of the biggest things I like to reiterate about any company is, especially on the stock markets, if you're in the mining business, you've got one or two options: you're either an exploration company, just looking straight for joint ventures or partnerships, or you're looking to put something in production. Either one of those situations is going to bring in cash flow. But, to have an actual controllable cash flow, you've really got to be looking at some level of production."

Lee glances out of the window and sits back in his chair. "That, to me, is the thing you've really got to be striving for, because that's what maintains your stock; that's what keeps your company strong. My feeling is that it shouldn't matter how big or how small the production is, if it brings in enough money a year in profit and it pays for your administrative costs, you're way further ahead than most companies that are out there." He concludes, "So, to me, that's got to be first and foremost. And, that's how you start protecting your shareholders' money and investments."

With an optimistic outlook, Molycor Gold's promising properties are exploration sites worth smiling about. However, alongside the good times, there have indisputably been tougher times too—a consequence of being a junior company in the mining industry for such a long period. "I guess that's where I'm a little fortunate," notes Lee. "I get to go along for the easier ride. "Larry had this company, you know, going through the hard times in the late 1990s, so he managed to survive all that, and keep the shareholders relatively happy, and keep moving along. The one thing we haven't done with Molycor is we've never rolled it back. At least while I've been here. And, the reason being, we've all got too much money invested in it."

With financing in place, Molycor is pushing ahead with their plans. An uplifting pace, in the eyes of Lee who has seen what can happen at other companies. "That was one of the things that Larry and I first did. That was our biggest discussion when we first met in 2003. It was, 'What do you want? You tell me what you want. If you want to do street promotion, don't waste my time. If you want to build a company, okay, then let's talk,'" recalls Lee. "I'm just not interested in street promotion. I could have easily kept my own office open and done that on the side. But, I just don't have time for it." Lee notes that he has no interest in wasting anybody's time. "Do it right, or don't do it at all," he says; another mantra by which he chooses to live.

Lee's refreshing attitude is likely a reflection of his years of experience in the business world. Yet, he sees no reason why others can't fall in-step with a more conscientious and straightforward approach. "Businesses change. I mean, the reality is that you've got baby boomers getting to the point where they're starting to retire. The last part of that boom, the fat part of the sausage as far as population is concerned, are guys who are my age, between the 40- and 50-year-old range. And these guys are starting to get into areas of decision making and they're starting to see things a little differently.

"You know what," Lee says, "you've got to treat everybody the same; you've got to treat everybody properly. And that means business is going to have to reflect that. We all have either fathers, or aunts or uncles that we know of, or grandmas or grandpas, that have put money into situations where companies have played musical chairs with their money and never given it a second thought." That is precisely the kind of irresponsible business Lee opposes.

A relatively small market cap company, Molycor is listed on the TSX-V (MOR), Frankfurt (M1V) and Pink Sheets (MLYFF). With some big prospects at Molycor, wise investors will likely be pondering the future of the metals market, and the stock market in general. Lee suggests, "I think the overall markets are going to take a little correction. I think they've gotten themselves a little too lofty. The bigger markets are going to correct, the smaller markets are going to correct, too, but I don't think by all that much."

He says, "Commodities are still in high demand and there's a really short supply of absolutely everything. You've got China trying to bring 230 million people into the middle class, well—good luck! I mean, for them to even get 10% of our standard of living, they're going to go through a whole year's, if not more, worth of consumables that the U.S. goes through, just to get them there, let alone maintain it. So, where are all these commodities going to come from, or resources? I don't know. I think the longer term is coming down to the haves and the have-nots. There are only so many resources around. And the squeeze is coming. And just sheer volume by numbers is starting to have its effect." A squeeze that may affect resource prices for many years to come.

Molycor's Ed Lee is aware of the trends and has the capability to handle whatever lies ahead in the markets and for Molycor in general. On the near horizon, he says, "With Crowrea, we'll be looking to try, like I said at the beginning, to put together a molybdenum resource. And with the Nevada package, well, you know—anything in Nevada, most of those companies go all the way. One drill hole makes a whole difference."

Lee concludes, "Certainly, it's hard to know what the upside's going to be, but we could start hitting with some of those properties down in Nevada. Yet, right now, I would definitely say that the moly property, Crowrea in BC, is the one that's going to be peaking all the interest. Everything's pointing in the right direction."

MOLY MINES LTD.

Reddened sands and rocky desert terrain stretch for miles. Australia's unforgiving outback is one of the most barren places on Earth. And it is here, in some of the oldest rocks on the planet, that Moly Mines Ltd. is developing their world-class molybdenum deposit. Long considered devoid of significant molybdenum, Australia is now poised to become a major producer of this important metal as Moly Mines begins development of their Spinifex Ridge property.

Moly Mines CEO and Managing Director, Dr. Derek Fisher, says, "Australia does not have a history of molybdenum mining. The deposit we have is a bit of an accident. It's different from most of the other deposits in the world, although it has similarities to the porphyry coppers, or the porphyry molys—the big deposits of the western United States, western Canada, and the Andes. But the rocks in this area of Australia are 3.4 billion years old, some of the oldest rock in the world, so it's a very different host environment than more typical porphyries are found in."

Spinifex Ridge is being touted as the first world-class molybdenum and copper project for the gigantic island continent. "Spinifex is certainly our focus. We have some other prospects in Australia, but nothing comparable. Australia's never had a significant moly mine. The moly mining that has taken place in Australia has been small. A lot of it was around the time of the wars. Now, there are several prospects happening down the east coast, but none of them have a large tonnage potential, like we have at Spinifex Ridge," notes Fisher.

Fisher began his career with the New South Wales Geological Survey in 1966 and, following graduation from the University of New England, Armidale, NSW, he moved into the mineral exploration and mining industry. Through his 37 years in the industry Fisher has established deep roots in both Australia and Canada. "My history in Australia goes right back to 1969, when I joined Inco. We were in Western Australia and I was involved in nickel sulphide projects, and then we were working up in the Solomon Islands. We were testing all sorts of things. We were working on nickel laterites, but we were looking for copper as well.

"I moved to Canada in 1973," Derek Fisher explains, during a recent Canadian visit to Toronto, Ontario. "I actually went to the University of Toronto, and, ever since, I've been associated with Canada, on and off. My wife is Canadian." Fisher undertook doctoral studies at the U. of T., earning his PhD in 1979.

Today, Fisher is mostly based in Australia for Moly Mines. But, the company does have an office in Toronto. "We've got one and a half people working there," he grins. "One full-time employee, Natalie Frame, who handles our Investor Relations, and we have another employee who's in three or four days a week. And then, in West Perth, Western Australia, we have our large head office of about 40 people.

"Personally, I have a house in Toronto, and a house in Australia; my kids are both Canadian and Australian. So, we tend to circulate a bit, since we have a base in both countries."

Moly Mines Ltd.—an independent Australian resources company—has a strong connection to Canada, as well. The company is listed on both the TSX and the ASX Australian Stock Exchange. "The stock is fully fungible in both markets. If you've got stock in the company and you trade it in Australia or here, there's no difference in the stock." In fact, Fisher has been the principal in listing a number of companies, both in Australia and Canada, and has more than twenty years experience as a manager of publicly-listed companies.

He has been around practically since the conception of Moly Mines. "Moly Mines was incorporated in January 2003 as Hibernia Gold Pty Ltd. "Hibernia was a private junior gold exploration company that had some properties in New South Wales. They asked me to come in to take the company public." Fisher joined Hibernia in February of 2003. "I spent about a year changing the company to a public one and getting the company listed.

"I actually had my eye on moly prices even then," Fisher admits. "We had a couple of moly prospects in NSW. I suspect that I was one of only a few people in Australia who had his eye on moly at the time. Don't forget, molybdenum prices had been flat for twenty-five years. Since the early 1980s, no one believed in moly. I didn't believe in moly, but, of course, we had these prospects; I had inherited them. So, I happened to be watching the moly price and, sure enough, within days of us listing on the ASX, in March 2004, the moly price went up—jumped from about $7 to $16," Fisher explains enthusiastically.

With moly prices fortuitously escalating, the Moly Mines CEO knew exactly where to turn next. "I knew this prospector who had Spinifex Ridge. It was, and still is, Australia's major molybdenum property.

"If you'd been around in the 1970s, you knew about Spinifex, because a lot of geologists in Western Australia had worked on it. It was discovered by Anglo American in 1969. It was drilled largely by Esso Minerals in the mid 1970s. And Esso controlled it into the 1980s. But just as they were bringing in potential partners for mine development, the world's moly scene collapsed. It was subsequently explored as a potential gold porphyry but it has no gold in it," Fisher laughs. "A few companies looked at it, but eventually they all walked away and a prospector picked it up. I acquired it from that prospector and, as a result, we changed the nature of the company almost immediately from a gold to a moly focus and changed the name, from Hibernia Gold to Moly Mines, to reflect that change."

Now the aptly-named Moly Mines is in a pole-position for developing a key Australian moly project that stands shoulder to shoulder with the big moly mines currently in development around the world.

Fisher smiles, noting, "My background is building mines. I'm one of these internationally-travelled characters. I built a copper operation in Mongolia in the 1990s. I was involved in copper in the Congo. I was one of the founders of the company Anvil Mining (the leading copper producer in the Democratic Republic of Congo), listed on the Toronto Stock Exchange. I was close to Equinox Copper Ventures, a big copper company (a 100% Zambian-owned subsidiary of TSX-listed Equinox Minerals). I was one of the founders of a company (Black Swan Gold Mines Ltd), which built two gold mines in Australia in the 1980s. I've been around," says the fifty-nine-year-old. "I think I've done the miles," he laughs.

As an experienced leader and businessman, Dr. Fisher seems best qualified to oversee Moly Mines' greatest molybdenum asset. Located in the 'development-friendly' Pilbara region of Western Australia, approximately 50 kilometres northeast of the town of Marble Bar and 140 kilometres east-southeast of Port Hedland; the Spinifex Ridge deposit was originally discovered more than 35 years ago.

Hibernia first optioned Spinifex from Kallenia Mines Pty Ltd., an arm's length party, on April 1, 2004. The agreement gave Hibernia a 5% interest in Spinifex Ridge, which could be increased to 90%. By February 10, 2006, the newly named Moly Mines exercised its option to acquire the additional 85% interest in the project through Moly Metals, one of its wholly owned and controlled subsidiaries. Kallenia retained a 10% free carried interest in the tenement.

Subsequently, Moly Metals acquired Kallenia's remaining interest, in exchange for 2.4 million shares of Moly Mines. Moly Metals now controls

100% of the Spinifex property and Kallenia retains a royalty payable by the company in the amount of A$0.02 per tonne of ore processed from the Spinifex Ridge deposit.

Meanwhile, back in 2004, the company undertook a scoping study technical assessment, using the historical data for the site. They concluded that the deposit should proceed to the pre-feasibility stage.

In 2005 and 2006, Moly Mines undertook a drilling program at Spinifex Ridge to confirm historical drill results and to better define the resource and its limits. Based on drilling conducted during 2005, a new independent resource estimate of 220,690,000 tonnes at 0.07% Mo and 0.10% copper indicated resources and 204,240,000 tonnes at 0.06% Mo and 0.09% copper inferred resources was reported at a 0.04% Mo cut-off. This resource formed the basis of the Spinifex Ridge pre-feasibility report released in February 2006.

In 2006, additional drilling work was required to increase knowledge of the grade, variability, metallurgy, and other key issues potentially impacting production rates and product specifications. This was completed near the end of June 2006, bringing Moly Mines drilling to a total of 198 drill holes for 36,641 metres. This data has been added to the already large database resulting from the previous work on the property, which included resource diamond drill holes; PQ sized diamond holes for a 20 tonnes bulk metallurgical sample; re-drilling of historical holes to increase hole depths; resource infill holes using RC drilling to depths of 250 metres; condemnation RC drilling to depths of 50 metres; infill RC drilling to depths of 300 metres; and geotechnical drill holes to gather information to estimate pit wall parameters. Ultimately, output from these drilling programs will be utilized in many aspects of the final feasibility study including detailed mine planning and scheduling, final metallurgical investigations, pilot plant test work, mill process design, and capital and operating cost estimates.

"Since we picked up Spinifex in 2004, we've spent over A$45 million dollars on the project. And, we've been steadily building up our staff over time. We are expanding our head office and we'll add 150 engineers from the consulting firm WorleyParsons. So, you know, we're moving ahead very quickly."

However, there was a brief delay that impeded completion of Spinifex Ridge's feasibility study. "We worked on the study for eighteen months, going on two years. We've had the opportunity to significantly expand the mining operation from 15 million tonnes per year throughput in the pre-feasibility study to 20 million tonnes per year." Fisher said.

When it comes to finding qualified workers to run the new mine, Fisher isn't worried. "There's nothing too complex in the mine itself," he says. "We're described as being a conventional mine with conventional processing techniques—it's a big open pit. It's not dissimilar to any big copper mine, but it's the first moly mine and processing operation of this (or any scale) built in Australia. There's nothing new in the metallurgy. It's metallurgy that's been around since the end of the First World War, when the world's first big moly mine went into production. That's one of the beauties of molybdenum; the mineral molybdenite has excellent metallurgical characteristics. So, there's nothing groundbreaking or anything in that.

"But, in terms of attracting staff, I've been in the industry for 38 years. And my COO/General Manager has been in it over 40 years. Therefore, we've got a lot of people behind us that we can draw on. And a lot of people have followed us into what we're doing; certainly in senior staff. We've had no trouble attracting people, and I don't think we'll have too much trouble, in terms of the overall mine. But, you know, it isn't easy, as you're probably aware, there's massive resource project expansion taking place in Australia at the moment. So, there's fierce competition for staff." And to stay one step ahead of that competition, as Fisher explains, "Our Human Resources Manager attended a recruitment convention in South Africa recently and we are advertising in South America as well. We have received an overwhelming level of interest in from skilled mining professionals wanting to be a part of the project."

Spinifex Ridge might seem like a world away from the First Nations claims commonly associated with mining in Canada, but Derek Fisher is cognizant of Australia's Aboriginal peoples and their rights. "Yes, they have Native title claims. We've completed all agreements with the Njamal people which is recognized in a Land Access Deed that provides employment, training and economic benefits to the Njamal people over the life of the project. All our mining leases over the mining areas are granted." Fisher explains, "You can't get your mining leases granted in Australia unless you settle these claims to the satisfaction of the government, and we've done that. That happened early in 2007, so that was a major milestone for the project."

"The next thing we have to resolve is our environmental permitting," says Fisher. "And that's in with the government at the moment. We're going through that process now, following about two and a half years of baseline studies we conducted in the area. There were no showstoppers identified, and we've also got the advantage that there's been a major mine, a big BHP iron ore mine, about 40 kilometres to the north of us. That was permitted only about two years ago, and it's in the same geology, the same landforms, and the same climatic area. Since that mine was successfully permitted, we don't foresee a problem with Spinifex."

Dr. Fisher elaborates, "In Western Australia, there are four levels of environmental assessment, depending on what the government considers to be the sensitivity of the area you're in. We're actually being assessed at the second lowest level. So, we're pretty comfortable, and we're not anticipating any problems from an environmental permitting point of view."

Anybody will recognize that Spinifex Ridge is no rainforest. Fisher laughs, "No, no, this is desert country. Hot, very hot desert country. Take for instance, the town of Marble Bar, only 50 kilometres away. It's an old gold mining town, and it holds a record and is well known in Australia. It's only a little town; it'd be lucky if there were 200 people in it. But Marble Bar has the record of having 180 days one summer where the maximum temperature was over 100 degrees. Every day for 180 days!" chuckles Fisher. "It's the hottest town in Australia. So, yes, we're in the Outback." But despite the remote and desert-like conditions, the location of Spinifex Ridge does have one gigantic plus. "We're only 150 kilometres from Australia's biggest tonnage port—Port Hedland. It's a big tonnage port because iron-ore is shipped from there. And iron ore is heavy."

Another boon to the location of this mine, on a more global scale, is it's proximity to the Asian markets. "Financing-wise, we're getting a lot of attention from major world banks as well as from the Asian steel industry, particularly China. Although China is a big moly producer—they produce 22% or so of the world's moly at the moment—China is increasingly consuming more and more of their own molybdenum. Certainly within the next five to eight years, they'll become a net importer of moly," proposes Fisher. And Spinifex Ridge is practically on their doorstep. "What you've got to understand is that where we are located is the Pilbara region of Western Australia—where Rio Tinto and BHP have their big iron ore mines, and their railroad systems and ports. It's where the Asian steel industry has been sourcing much of its raw material for steel for the last 40 or 50 years. And it's not just iron-ore, they also source manganese there now.

"So, in terms of being well located, being understood by the Asian steel industry; you couldn't be in a better location," says Fisher. Not surprisingly, Moly Mines is open to joint ventures should they arise in the future. "Yes, we're getting a lot of attention, particularly from the steel industry in the countries surrounding China; countries which traditionally sourced their moly out of China. But as China uses more and more of its store of moly internally, the steel industry in the circum-China countries is getting nervous. Because we're so advanced with the project the major Chinese steel companies are starting to knock on our door. And Asia has been a big investor in the Pilbara region in the past as well. Of course, from our mine development perspective, you couldn't get a much better location—no question," says Fisher.

"We've got timelines, we've got a complex management program for securing financing, for developing the project," explains Fisher. "We're aiming to have financing in place by early to mid first quarter of 2008. We've got an advisory group out of Australia called Azure Capital; they're assisting us in terms of running parallel timelines on everything from debt financing to equity financing to high-yield bond type financing to cornerstone investors. We're running parallel schedules for different styles of financing and that is coming together as we get closer."

This forethought and diligence is necessary when building a new moly-copper mine with a lifespan exceeding 20 years. "The Spinifex Ridge has, at today's metal prices, contained metal of somewhere between 25 and 30 billion Australian dollars," says Fisher. "One of the important factors is that it does have some copper, and copper will assist us in banking the project, because there's a forward market for copper, you can hedge it and you can do things with it. You can't do those things with moly, so that's one of the sweeteners we had with this project, in terms of financing it."

Fisher smiles, "It's a big resource. Very big. With sustained high moly and copper prices, and the expanding resource we're looking at now, and the possibility that we actually have the ability to expand this project to about 27–30 million tonnes per annum – this is really significant. At those throughput rates, with a cut-off grade of 0.02% Mo, the tonnage in the project will probably be greater than a billion tonnes. So, the project demands scale. And we haven't drilled to the end of it. We're actually going to do that in the near future. Our intent is to launch another drilling program to better define the limits of the deposit." Ultimately, he says, "This will be a long-life mine."

Spinifex Ridge will not only be a long-life mine; it will be a unique mine in Australia. "We're a bit of an accident of nature actually, because we're in some of the oldest known rocks in the world," says Fisher. "There's only a couple of places in the world where you get rocks of this age, and most modern moly deposits appear in very young environments—like along the

Rocky mountains and the Andes. We don't fully understand why Spinifex is here—and we don't really care why," he laughs.

Australia is by no means a known player in moly. In fact, the distribution of molybdenum reserves and production is essentially concentrated in only a few countries in the world. China, USA, Chile, and Canada hold nearly 90% of all known moly deposits. The USA, Chile, and China are the main producers, sharing 75% of output with each producing approximately 30,000 tonnes per year. Nevertheless, based on the success of Spinifex, Moly Mines is exploring several other highly prospective molybdenum projects within Australia.

In the state of New South Wales on the east coast, the company has acquired the Glen Eden, Mt. Pleasant, and Mt. Tennyson properties. "We've got some prospects in NSW," says Fisher. "They're exploration targets. We will work on them, but you've got to understand, to put it in context, the Spinifex Ridge is such a big resource, and it's a resource that we really haven't got to the end of."

In February 2007, Moly Mines announced that it had entered into a binding term sheet with ASX-listed Cortona Resources Limited to sell its NSW gold assets—remnants of the company's gold exploring beginnings—in a cash, share and royalty deal.

Yet, Fisher confirms that the company is truly occupied with Spinifex Ridge. "We appear to be the most advanced of the big new projects out there in the world today. We're probably twelve to eighteen months ahead of our peers. So, this project, from that point of view, has a very high likelihood of being financed, and moving to development."

As a result, Moly Mines is something of a trailblazer on the continent. There are now several moly explorers who are hoping to follow in the footsteps of the Australian company. "We prompted a few of them," states Fisher. "There are other people following us now. Australia does have other molybdenum prospects around, but none of them are showing any signs of being of the scale that we have here." Moly Mines has a genuine edge over any rivals, not only because of the scale of the Spinifex deposit, but because they are over a year ahead of any contenders. Fisher highlights, "You know, we've ordered our long-lead time items for the mine already. We did that back in April 2007.

"In April, we signed an agreement for the primary crusher and tertiary crushers, and ball mills. That was quite a coup for us because they actually put us on the order list as of January 1st 2007, even though we didn't sign the deal until the end of April. Delivery time is twenty-four months from

January 1st . But, if you were to order those items today, delivery is now thirty-six months," notes Dr. Fisher. "Basically, you can't get the stuff. You know, that's one of the big advantages we have at the moment; we are ahead of the pack. Trying to order that sort of gear now, it's not going to be easy for other companies."

It won't be easy for other companies to compete with Moly Mines' impressive management team either. "Collis Thorp, our COO/General Manager, is a 40-year mining veteran. He built and managed mines both in Australia and overseas. He has deep experience in that, and he's been running and managing the feasibility study at Spinifex Ridge." Thorp's mine experience covers precious, base, and specialty metal operations including gold, tin, tungsten, titanium, and zircon. He has also served as Managing Director of the Cable Sands Group which operated four mineral sands mines for titanium and zircon in Western Australia.

The rest of the team is equally qualified. John McEvoy, Chief Financial Officer, has extensive experience in senior finance roles in both the public and private sector and approximately 15 years experience in the mining industry. Andrew Worland, Company Secretary and Manager–Commercial and Marketing, has been significantly involved in all financing and corporate transactions and investor relations activities since the company listed on the ASX. Mike Gloyne, Manager–Mining, is a mining engineer with 25 years of mine experience covering a range of commodities including iron ore, gold, coal, nickel, and mineral sands. David Pass, BSc, Manager–Metallurgy, earned a double major in Extractive Metallurgy and Chemistry, and brings more than 20 years of mining industry experience in mineral processing plants to the table. Graeme Kininmonth, Manager–Health and Safety, and Environment, is a 20-year resources industry veteran. Jeff Reilly, Manager–Marketing, has 25 years of international sales and marketing experience developed during senior sales and marketing roles within the industrial minerals sector. Brendan Cummins, Exploration Manager–Molybdenum Projects, graduated in 1993 from UWA with First Class Honours and has been working as a geologist for the past 15 years, including supervising the drilling and resource estimation of the Spinifex Ridge resource. and Barbara Kwiecien, Manager–Land Administration is a geologist with extensive experience in Canada and Australia both in mining and land management and native issues both as a negotiator and a mediator.

Paul Willis is Chairman of Moly Mines' Board of Directors. He has extensive project analysis and capital markets experience. He has been responsible for managing portfolios of listed Australian and international resource companies across all resource sectors, particularly gold, base metals, bulk commodities and oil and gas. "Our former chairman was Andrew Forrest," adds Fisher. "He approached me with respect to creating this company. Andrew is a famous Australian entrepreneur; probably the most famous." His personal net worth is valued at A$5.0 billion, and he was listed on Forbes' 2007 richest people in Australia and New Zealand. Forrest founded the Fortescue Metals Group, which is developing an iron mine in Western Australia. "It's a big iron ore deposit, not far from where we are. And he's building a 280-kilometre railway line, he's building a port, and he's building a very large mine. He still owns somewhere around 15% of our company.

"So, as you can see, we've got a very good crew around us," says Derek Fisher.

Together, the Moly Mines team is focused on developing Australia's first world-class molybdenum and copper property. "You know, we're trying to do things in parallel, not in series, and there is an opportunity out there. While molybdenum prices stay high, the opportunity is there for a junior company with a big project like this." Moly Mines appears, in every way, to be that company.

MOSQUITO CONSOLIDATED GOLD MINES

To the casual observer, their name—Mosquito Consolidated Gold Mines—might imply a gold-centric outlook on mining opportunities. However, as CEO and President Brian McClay explains, Mosquito is now more about moly than ever before. "Moly is now the primary focus of the company," says McClay. "I mean, you're trying to develop a mining company and it's a trip down the yellow brick road—who knows where it's going to end up, or what the stops are along the way." It's been a long road for this Canadian natural resource Exploration and Development Company—but one that has led them to a diverse portfolio of quality precious- and base-metal projects across North America and in Australia, including the Idaho-based Cumo Molybdenum Project.

Located in the mountains of south-central Idaho, the Cumo Molybdenum Project is a potentially incredible moly resource expected to put Mosquito permanently on the map. Moreover, the company predicts that Cumo's by-products of copper, rhenium, gallium and silver will help to offset any production costs for moly—ultimately making Cumo one of the most profitable molybdenum mines on the planet. Cumo, therefore, leads the pack of Mosquito's promising properties.

In Canada and the U.S., Mosquito's other assets include a gold properties portfolio consisting of Cariboo, Brett, Statlu, Red Lake/Laverty, Larder Lake/Cummins, Black Point, and moly properties that include Spring Creek, Spruce Mountain, and Pine Tree. "Pine Tree is slightly different from Cumo," points out McClay. A molybdenum project, situated in the Pilot Mountains of western Nevada, the Pine Tree is located within a 65-mile belt that hosts numerous gold, silver, and copper deposits and mines known as the "Walker Lane Trend." The primary target on this property is a 500 million ton copper–molybdenum–rhenium porphyry deposit with potential gold, silver, indium, and gallium by-products.

Naturally, among these various properties, McClay says, "Although moly is very hot, we also have excellent gold assets. The Trikay project, for example, a extensive and undrilled copper-gold mineralization in Queensland, Australia. We're just responding to opportunities as they come along. And in Nevada we think we have a major Carlin Gold-type shot with our Black Point property. Now that deposit could be worth, potentially, a billion dollars, or it could be worth nothing. So, you're betting three or four hundred thousand dollars to win a billion. Obviously, we want to take that shot."

McClay confides, "I've found that the formula for success in the mining business is pretty simple: you have to have the projects, you have to have the people to pull it off, and, of course, you have to have the money. "A few years ago," he continues, "the money wasn't available. You'd get the projects and the people, but no money. Now the money is available but the good deposits are tied up and obviously good people are very tough to find. As we're trying to be a world champion junior mining company, we're following that formula. I think we've got a first class technical team, we've got the makings of two world-class moly deposits and we have over $10 million in the bank. The Cumo is the more advanced. It's larger, the dynamics are good, and it could be among the world's lowest cost producers."

With investments in several countries, McClay highlights that low political risk is one of the essential features Mosquito looks for in a project, based on lessons learned in the past. "The knowledge that we've acquired," he says. "when the yellow brick road took us to Russia for six or seven years. Basically, I spent the prime years of my life in a frustrating exercise in Russia. There's such a high political risk over there, especially in the gold business. Gold is very political in Russia and in the Third world. Moly, base metals, they don't quite understand them, so they don't care. But, if you've got a major gold deposit, you're probably faced with hostile political forces forming against you to either take it from you or tax all the profit from it.

"Who wants to put a decade's worth of their life's work in the hands of some politician that doesn't care about you? So I'm now going to leave that to somebody else. We gave up a lot in the search in Russia. I'm not into chasing rainbows." McClay concludes, "I find that pioneers usually get arrows in the ass. And, years later, am I any wiser or older? Well, either way, we now focus our search in low risk, politically secure, low cost exploration environments."

Along with harsh lessons learned, Mosquito Consolidated Gold Mines has weathered some seriously dismal market conditions—one of the consequences of being in the mining business for many years. "And we got through it all without diluting the hell out of the company!" emphasizes McClay. "You keep in mind that I came in here with a chair and a whip for a long time, trying to ward off dilution; generating cash internally." With a colourful flair for metaphors, he adds, "So, here we are—we've toiled in the vineyard and hopefully we'll get to pick some fruit."

Toiled indeed. McClay is an example of a hard worker who has personally toiled most of his working life in the mining industry, starting as a miner in the family business. "You know, the mining business as a rule is very risky. It's a tough, tough business. And it's a town without pity, the mining world. When you drill a property, I tell people, five things can happen, four of them bad. Especially a gold property. I've been waiting most of my life to find a strong base metal market. Because, in these big porphyry deposits the mineralization is much more predictable, more consistent, and easier to drill."

Born and raised in Vancouver, the sixty-one-year-old McClay explains, "When I was a young man, I actually worked at the Highmont Mine for four or five years, during the development phase. It ended up being among the largest mines in Canada. My dad was one of the founding Directors." McClay nods philosophically, "My family were open-pit mining contractors and I also worked underground in mines"—something that not many Presidents of mining companies can attest to. "And I've been a diamond drilling contractor, so I have experience in many facets of this business."

In total, McClay has amassed over 40 years of experience in the international mining industry—prospecting, contracting for mineral exploration, diamond drilling, open-pit mining, and managing open-pit and underground operations. Consequently, over time, he has developed a strong reputation as an intuitive mine builder. For the past decade and a half, McClay has directed Mosquito Consolidated Gold Mines Limited, as President and CEO from 1990 to 2000 and again from 2004 to the present. A master negotiator and deal-maker, his combined field and management experience are unmistakable assets to the company as it embarks on a new age of growth, answering the intense global demand for molybdenum and other metal resources.

"Absolutely, moly is what we're focused on now. These are world-class moly deposits, Cumo especially, there's no question. We think it stacks up with the best in the world, if not the best," states McClay with confidence. "And what makes Cumo even more attractive is our discovery of electronic minerals—rhenium, gallium, and indium. I think they carry the whole mining cost of the project – so the moly, silver and copper are almost free."

Mosquito Consolidated was founded back in 1971 as Mosquito Creek Gold Mines Ltd, and put the Mosquito Mine into production in 1980.

Mosquito's current management, however, first took the reins in 1991, which included taking responsibility for the high-potential Mosquito Creek claim group in eastern British Columbia. "We have been actively exploring, globally, for over 15 years," says McClay, noting that he feels that their market cap is "grossly undervalued" at current levels.

Mosquito acquired the rights to the Cumo claim group—344 claims covering an area of 2,867 hectares—from Cumo Molybdenum Mining Inc., in the fall of 2004. The Cumo property already possesses great infrastructure, which includes a solid, all-weather highway and logging roads providing year-round access from Idaho City, located just 15 miles southwest of the property.

Due to a depressed metals market, Mosquito obtained a favorable price for the property, despite its large potential.

Cumo was originally discovered by Midwest Oil and Gas, which was subsequently acquired by AMOCO in the late 1960's. Cumo was explored by Amax-Climax Molybdenum and AMOCO Minerals who completed multiple stages of exploration between 1968 and 1982. This exploration included geological mapping, geochemical surveys, geophysical surveys, 36,022 feet of diamond drilling and a computer-generated block model geologic reserve calculation. As indicated by Amax's historical 1982 block model, Cumo is thought to be host to a geological resource of 1.387 billion tons of 0.093% MoS2 at a 0.05% MoS2 cut-off. This resource includes a higher-grade core of 444 million tons of 0.135% MoS2 at a 0.10% MoS2 cut-off.

With Amax's historical resource in hand, McClay comments, "Amax says we have 1.387 billion tons of moly ore, as well as a 2.2 billion ton resource, and a 450 million ton high grade core. We also think they got the model wrong. Our last drill holes show the moly content substantially higher. Therefore, we think it's a much, much bigger deposit than two billion tons. Possibly double or triple the size. But, that being said, let's call it 1.387 billion tons. If you take that figure and multiply it by Amax's grade; this still makes Cumo the fourth largest contained moly deposit discovered in the history of man."

The vast size should spell a lengthy lifespan for Cumo, which also makes it an outstanding prospect for Mosquito. "the large multi-national smelter companies want to see long-term reserves. When I say long-term, I mean 50 to 100 years. That way they know it can be mined through multiple price-cycle peaks. And they're only going to have to permit it once. They can also establish the most efficient infrastructure available. This all equates to being among the lowest cost producers in the world." McClay adds, "Most importantly, when you have a deposit the size of Cumo, the production rate is as high as you want—you get the largest truck fleet available and in the mining business, the bigger the truck, the lower the mining cost. It's just that simple. While these other guys, who don't have a long term deposit, they're stuck with 150-ton trucks. But we're going to start rolling down the road with 450-ton trucks, you know, that equates to dropping our mining costs from $12 or $15 per ton to around $5 per ton. That model proves that we can be in the lowest-cost percentile. It's not magic."

Of course, McClay adds, "our task at hand is to complete a 43-101 resource for the deposit. Then it's no longer Brian McClay sitting here, saying, 'Oh, we've got over two billion tons.' We need about five holes for the 43-101, and we've done two already. Of the last two holes we drilled, each 2,500 feet deep, one bottomed in ore. So, we're going to finish these holes, then complete a 43-101 resource in the winter of 2007. Along with that we'll do a scoping study, which will give a cash-flow analysis; an economic evaluation; and a net present value for a financial appraisal of the project. And those numbers—you can divide the net present value by the amount of shares out—that number's going to stagger people, I feel.

"Once the numbers are in you drop back, you evaluate, digest. Obviously, we have had interest from major mining companies, so now the next step is to finance the project properly and to drill it out. I mean, I'm not drilling this just to get a 43-101 reserve; I'm trying to delineate the size of the deposit and the grade. That's my role in all of this, to exploit it for as much as we can, for the benefit of the shareholders."

As McClay follows his mining formula to its inevitable conclusion, he asks himself, "Do we have the deposit? Yes. Do we have the money to pull this off? Yes. Do we have the technical team? Well, I think I've got one of the best geologists in the world running this thing. You can't do this with dummies. I've been at this for 40 years; I've met some 3,000 geologists—and they run the whole spectrum from moron to genius. This is technically demanding, and young geologists just don't have that experience to develop these large scale deposits to the standards required today."

Dr. Matt Ball, PhD, PGeo, is the Senior Geologist for Mosquito Consolidated, and a Director of the company. Ball is a member of the Association of Professional Engineers and Geoscientists of British Columbia, the Society of Economic Geologists, U.S.A., and has 25 years international experience in exploration and operations geology with chief geologist positions at such diverse locations as Erickson Gold Mines in B.C.'s Cassiar region and BHP's Ok Tedi Mine in Papua New Guinea He was a semi-finalist winner of the Goldcorp Challenge—and was subsequently employed by Goldcorp—and he subsequently worked for the overall winner, Geoinfomatics Explorations Inc.

In early 2007, Ball was busy sorting and organizing over 2,000 sample-rejects as part of a re-assaying program from past work on Cumo, with Exploration Manager and Director of Mosquito Shaun Dykes, MSc (Eng), PGeo. Prior to joining Mosquito Consolidated in 1994, Dykes spent 15 years with Westmin Mines Ltd of Vancouver, as a Senior Project Geologist. "You know, Shaun is absolutely essential," McClay states, "he's the geological brains behind this thing. Mosquito isn't just about Brian McClay, obviously. I'm the Captain of this ship, but it doesn't go anywhere without a good navigator."

Mosquito has obtained original rejects from the drilling completed between 1976 and 1981 by AMOCO and Standard Oil (Indiana), and Amax and Climax Molybdenum Company. The drilling in that period consisted of a total of 33,396 feet in 23 diamond drill holes and 2,630 feet in three air rotary holes. Initially, Mosquito sent one out of every ten of these samples for check assays, to confirm the grades reported by Amax. In addition, the samples were analyzed for rhenium, gallium, and indium, which were not previously assayed. These metals, along with copper and silver, could prove extremely significant in sealing Mosquito's ambition to turn Cumo into a low cost molybdenum producer.

As of May 2007, 176 rejects were sent to ALS Chemex of Elko, Nevada for re-assaying. The samples were analyzed for a total of 45 elements. Along with excellent correlation coefficient numbers between the original molybdenum and copper assay results and the new Mosquito results at Cumo, significant quantities of rhenium, gallium, and silver were confirmed. Rhenium ranged from 20 to 78 ppb, averaging 43 ppb in the molybdenum zone; gallium ranged from 10 to 26.5 ppm, averaging 17.9 ppm; and silver ranged from 0.07 to 26.3 ppm, averaging 2.8 ppm.

Overall, these are the kinds of figures that get investors' tongues wagging, and for Mosquito President Brian McClay, this is only the beginning of something really huge. "We've got enough money to take it to the next stage, and I feel that next stage is the low risk stage. The drilling to date has displayed the consistent mineralization found in all the great deposits of the world. We think it as close to an arcade duck shoot as you can find."

So, as Mosquito Consolidated takes aim at some potentially valuable moly targets at Cumo, along with a host of other interesting metals; McClay next outlines the plan for their Nevada-based Pine Tree property. "The Pine Tree, we think, is another great deposit," notes McClay, "but, we haven't penetrated the core; usually these porphyry deposits have a high-grade core. We're migrating towards that. That is our drilling ambition—the best is yet to come. We know our intersections are getting thicker and richer in every hole. So, our geological model is confirmed and we've got a good structural picture; it just takes some more drilling."

The first of Mosquito's 25-hole summer 2007 drill program at Pine Tree encountered significant copper–moly mineralization. The 383.9 meter intersection graded 1.01% copper equivalent and 0.090% molybdenum equivalent. Copper and molybdenum equivalents were calculated due to the multi-elemental nature of the mineralization. The hole was designed to target a thick molybdenum bearing zone intersected in another hole to the north. The hole intersected the thickest section of molybdenum-bearing mineralization and an associated intrusive quartz–feldspar porphyry, confirming an established geological model of increasing grade and thickness to the north of the property. Mineralization was found to consist of molybdenum, copper, gold, silver, rhenium, indium, and gallium.

"Pine Tree was developed by FMC and all sorts of other guys," McClay explains, "other companies that were looking for a copper–gold deposit," notes McClay. "Everybody's looking for gold in Nevada and they didn't figure out the moly model at Pine Tree. They just looked at the mineralization and chased that mineralization instead of a larger geological picture. Most often commodity economics guide geological expeditions."

The Pine Tree deposit was originally discovered in the 1950s and was jointly owned by Continental Mining and Bear Creek Mining. The property was subsequently re-staked by Nevada geologist Tom Evans and prospector Jim Meyers. In May 2005, Mosquito Consolidated Gold Mines Ltd. entered into an option-to-purchase agreement with Evans and Meyers, acquiring a 100% interest in Pine Tree, which is anticipated to be an exceptionally worthy copper–molybdenum resource.

Molybdenum is paramount for Mosquito, as evidenced not only by the moly potential at Pine Tree and Cumo, but also by its recently staked Spring Creek claims. The claims are a two-hour drive from Cumo and boast a series of molybdenum–silver–copper bearing quartz veins. Values from underground and surface chip samples range from 0.20% to 1.64% MoS2 and from 17 g/t to 68.6 g/t silver. An aggressive exploration program is planned on this property as soon as human resources become available.

Yet, Mosquito Consolidated Gold Mines still has deep roots in gold mining. Its other current projects include its Mosquito Creek Mine and Cariboo Gold Quartz Mine in the famous Cariboo Mining District of British Columbia. Presently, 100% of Mosquito Creek and 50% of Cariboo Gold Quartz Mine are under option to International Wayside Gold Mines Ltd. "So, we have some royalty holdings; the Mosquito Mine, I believe, is the most advanced gold deposit in western Canada, not already in production." says McClay.

Additionally, Mosquito has a joint venture interest in the Brett gold exploration property in the Vernon, B.C. area; gold exploration properties in the Red Lake and Larder Lake areas of northwestern Ontario; the Statlu gravel near Vancouver, BC; and the Black Point gold exploration property in Nevada; as well as the Trikay copper-gold exploration property in Australia. "You know, those are part of the foundation of the company" affirms McClay. "So, yes, there are many dimensions to this company and other dimensions to be."

He leans forward, "I personally think the stage is being set for a major gold price and it's going to be linked to political events more than economic, and that political event is the U.S. Presidential election. You know, the U. S. economy is like an ocean-liner, it takes forever to turn around, it's very slow. Nothing happens dramatically. Anyway, President Bush, at the end of his term, is going to want to suppress any bad news. He wants to get a Republican elected, whatever it takes—money supply, low interest rates—whatever it takes to keep the economy as buoyant as they possibly can, that's going to happen. Conversely, the new President is going to want all the bad news out of the way in the front end of the term, and that's where you're going to see, I think, economic carnage; which is going to result in a spectacular gold price."

Speaking about several of Mosquito's gold properties, McClay explains, "Right now, we own these assets 100%, we don't have any payments to anybody; they're just sitting there. Our technical team is focused on developing the moly properties—particularly the Cumo. You know, we're not a major mining company. We don't have the resources, so we have to stay focused. And many, many things can go wrong and you can't be a man who chases two rabbits and catches none. "You know, I'm a miner. I've worked around mines most my life, and old miners say: 'Stay in the ore, stay in the best ore.' So, just keep focused."

McClay continues, "Not forgetting that, when you drill a gold property, even the best gold vein, they only mine 25 percent of it, only 25 percent of it carries, so there's a lot of hit and miss in the gold, whereas these big base metal mines, or these big porphyry systems, are more predictable." With his mind clearly focused on the task in hand, McClay looks ahead, noting, "As markets are all about news and momentum, we're getting to the stage right now where, to keep the drilling going – you want to be a drill hole promoter. The market is going to react to positive drill hole results and basically not much else. There are red letter days ahead, like the release of the 43-101 resource, but it will take drilling and we're on it."

To feed Mosquito's dedication to keep drills turning on their projects in a marketplace where drilling contractors can be difficult to come by, the company recently made a bold move. They bought a drilling company. Mosquito Consolidated started out by buying a few drills, but eventually acquired a 100% interest in Kirkness Diamond Drilling Inc., of Carson City, Nevada, for $3 million (USD). The drill fleet of Kirkness includes both underground and surface drills with a capability of supplying a wide range of drilling services to the mineral exploration industry. Also, Mosquito will incorporate its own drills into the Kirkness drill fleet, including an EF-50 core diamond drill, an air rotary/reverse circulation drill and a new Atlas Copco CS-14 core drill.

The combined Mosquito/Kirkness drill fleet will be able to supply all of the drilling requirements of Mosquito in the foreseeable future. Kirkness will also continue to operate as a drilling contractor in the western United States.

In addition, other big moves in Mosquito's future might include handing off the Cumo, if the situation suits. "There's this business of taking the ball down the field as far as you can, before you have to pass it off to somebody else—a major. Obviously, the further down the field you get, the more the balls are worth," says McClay with a smile.

So, will Mosquito pass off the Cumo eventually? "Like I said, it's a trip down a yellow brick road," Brian McClay says. "I'm at the age that I don't know what will happen, or how far we can get. We've attracted interest from major smelter groups and, you know, they want to make their money. They can make their profit out of the smelting end of things; leave the mining to the miners. But, Cumo will require capital expenditures of well in excess of a billion dollars, perhaps approaching two billion dollars, so it's getting beyond the range of a company the size of Mosquito, but who knows. I'm not saying we're trying to sell it; we'll see what happens."

As for acquiring additional properties to add to an already wide portfolio by comparison to many juniors, McClay says, reaching for another apt analogy, "This isn't like sport fishing, where there's a catch limit. You know, we've got our hook in the water; things come around." Of course, often that hook is carefully placed in the right waters to catch that prized property. "I mean, look at the portfolio of properties we have. This didn't happen by accident!" laughs McClay. "Some of them we got for next to nothing, because we recognized something that wasn't previously seen."

McClay explains his secrets to success in this business, "Do you have to be smart in this business to win? Absolutely. Do you have to be cunning? Absolutely. Ruthless? No. You just have to be diligent, hard-working, keep focused, and show perseverance—almost to the point of obsession." McClay and Mosquito's dogged perseverance has undoubtedly led to their lengthy and beneficial terms in the mining business. "I'm old enough, believe me. I wish I was younger, I wish this metal market happened earlier in my career," says McClay. "There was a serious crawl through the desert. There was almost a holocaust in mining and I knew we didn't have any money; we had very little money. And it's all about perseverance. It's like a card game; you've got to have patience and perseverance."

With his patience beginning to pay off, McClay notes, "Of course, we have some excellent gold assets that seem to have been lost in the shuffle in comparison to these moly projects. But, my plan is to eventually spin out a gold company, or spin out the moly—one or the other." Ultimately, McClay asserts, "Our properties are all low political risk and low environmental risk. I think we've got the upside of all these electronic and technical minerals. These are strategic minerals of interest to the United States, and that's a big deal. I think moly prices are going to correct. But, are rhenium, gallium, and indium going to go down? No, I think the demand is rising..

"That is the key—this multi-elemental mix, so you're not vulnerable to the collapse of the price of just one metal. Cumo has a major silver component in it. We know there's a gold zone at Pine Tree, we haven't had the time to drill and chase it down; what we're chasing now is the high-grade zone and confirmation of a much larger and different deposit that was modeled by Amax, and we've basically got it all figured out— the next few holes will demonstrate that. Diverse minerals mean that we're not so vulnerable to market fluctuations," concludes Mosquito Consolidated's Brian McClay. With what's going to happen in the next three or four years with metal prices, these diverse electronic metals, combined with two world-class molybdenum properties, are going to make us stand out above everybody else."

PACIFIC BOOKER MINERALS INC.

All eyes are on the Morrison porphyry copper–gold–molybdenum deposit in central British Columbia that is currently in the advanced stage of development by Pacific Booker Minerals Inc. Located in central British Columbia approximately 35 kilometres north of the village of Granisle, the Morrison is now on course with its Feasibility Study and Environmental Assessment, conducted by Wardrop Engineering Ltd. and Rescan Environmental Services, and expected to be completed in the fourth quarter of 2007. Pacific Booker is proposing an open-pit mining and milling operation for the production of copper, gold, and molybdenum concentrate.

To date, a total of 29,009 metres of core drilling (125 holes) have been completed as well as a series of geotechnical studies, metallurgical testing, mine layout and open pit design, tailings studies, environmental base line studies, and socio-economic studies. The measured/indicated mineral resource is 206,869,000 tonnes grading 0.46% Cu equivalent. In addition, there is an inferred resource of 56,534,000 tonnes grading 0.47% Cu equivalent.

The Morrison Project is within 29 kilometres of two former producing mines, Bell and Granisle, consequently providing the Morrison with the advantage of an existing regional infrastructure, including a deep-sea shipping terminal at the port of Stewart, B.C., a road network, nearby power (25 kilometres from project site), and a full service town (the Village of Granisle) within daily commuting distance from the Project site.

Executive Vice President and Chief Operating Officer of the company, Erik A. Tornquist says, "I think our advantage is that we already have a proven resource," relying on the recently announced NI 43-101 report for their wholly-owned Morrison property, that is spurring them on, and attracting welcome attention from investors; relatively straightforward metallurgy with good recoveries and acceptable concentrates, an existing regional infrastructure, and the support of the local communities.

Finnish born Tornquist moved to Canada from Sweden with his family at the age of ten. An Applied Science Technologist with over 30 years experience in Natural Gas Operations, Engineering, International Project Management, Human Resources, and Training, he has held various management positions with Terasen Inc.—the largest distributor of natural gas in British Columbia—and Terasen International. Tornquist began with Booker as an investor, joining the Board of Directors in 2005 at the request of Executive Director John Plourde. "I was a shareholder," Tornquist explains, "and I was getting frustrated at Pacific Booker's lack of focus." During a restructuring that involved replacing five of the companies seven Directors, "John asked me to come on board," Tornquist says, "and here I am.

"Whether working with natural gas or minerals there are a lot of synergies. My background is instrumentation, systems and automation which is a multi-disciplinary science, and project management. I worked for 10 years directing international projects. So, there are a lot of synergies." Most recently, Tornquist was overseas in the Sultanate of Oman, southwest Asia. He worked as Vice President of Human Resources and Training for Canadian Energy Services that had a contract with Oman Gas Company to operate a 1000-kilometre natural gas transmission pipeline and commission a second one. "We started a company from scratch, so we had to staff the company, train employees and write all the policies and procedures, operating instructions, set up all the quality control programs, etc."

Moreover, Tornquist has completed PUBCO 1 at Simon Fraser University, a course on setting-up and managing public companies; ideal considering that Pacific Booker is now listed on two North American exchanges: the TSX-Venture (symbol: BKM) and, as of August 8, 2007, Amex—the American Stock Exchange (symbol: PBM.). The Amex listing is the latest move for the company with a history dating back to 1983, when it was originally incorporated as Booker Gold Explorations Limited.

In 1992, Booker Gold started to work on the Hearne Hill claims, a property adjacent to Morrison and famous for the discovery of the $100,000 boulder—a large copper, silver and gold laden mineralized boulder uncovered by a bulldozer in the 1960s. In 1997 Pacific Booker Minerals Inc. entered into an option agreement with Noranda Inc. to obtain a 50% interest in the Morrison property. Pacific Booker subsequently purchased the Morrison property from Falconbridge Limited (formerly Noranda Inc.) in April 2004, with no net smelter return or concentrate commitments to Falconbridge Limited. Booker Gold consolidated its stock in 2000 and changed its name at that time to Pacific Booker Minerals Inc.

"When the company started to develop the Morrison, they ceased working on the Hearne Hill claims," points out Tornquist. "They drilled

143 holes totaling 33,493 metres, on Hearne Hill." Drilling outlined a high-grade copper–gold deposit, "however it's very low tonnage," Tornquist explains. "When the management at Pacific Booker was changed in 2005, we decided to review the merits of Hearne Hill and it was deemed to be uneconomic."

The Morrison copper-gold porphyry deposit was discovered in 1962 by Noranda Exploration Company. Noranda completed 95 exploration drill holes totaling 13,893 metres from 1963 to 1973 that broadly defined the deposit to an approximate depth of 150 metres. Noranda published two resource estimates for the Morrison Deposit. The first was 86 million tonnes grading 0.42% copper. The second estimate calculated in 1992 gave an inferred resource of 190 million tonnes of 0.40% copper and 0.21g/t gold. No further drilling was done until Pacific Booker Minerals optioned the property in October 1997. Pacific Booker conducted a National Instrument 43-101 compliant, three-phase drilling program, which commenced in January 1998 with the primary objective of defining the configuration and depth of mineralization for the deposit, determine the geological controls for the mineralization, confirm copper and gold distribution, and determine if the project could be advanced to the feasibility stage. Since 1998, Pacific Booker Minerals completed 96 exploration core holes on the Morrison deposit totaling 26,202 metres. Four additional large-diameter PQ core holes were completed in 2005 for metallurgical test samples and 25 holes were drilled in 2006 for geo-technical and site evaluation studies.

The results provided Pacific Booker with a substantial copper–gold–molybdenum resource. With a measured and indicated mineral resource of 206,869,000 tonnes, grading 0.46% copper equivalent, the Morrison resource consists of 0.39% copper, 0.20 g/t gold, and 0.005% Mo. The contained metal is 1.8 billion pounds of copper, 1.3 million ounces of gold, and 20 million pounds of molybdenum. In addition, there is a sizeable 'inferred' resource of 56,524,000 tonnes, grading 0.47% copper equivalent, consisting of 0.40% copper, 0.21 g/t gold, and 0.005% Mo.

Metallurgical test-work conducted by Process Research Associates Ltd. (PRA) and IME Consultants Inc. (IME) indicates that the metallurgy of the Morrison deposit is relatively straightforward and that good copper recoveries and acceptable concentrates can be achieved. PRA estimates 85.7% copper and 59.7% gold, and IME estimates a slightly higher 88.4% for copper and 63.0% for gold. SGS Canada recently completed a Feasibility level metallurgical test program and the final report is expected at the end of the year. The Metallurgical test-work program conducted by SGS

Canada will also include recoveries for molybdenum. It is expected that metallurgical recoveries would be very similar to the recoveries achieved at the nearby Bell Mine.

The proposed mine is being planned as open-pit with a primary crusher, plus a waste and tailings disposal site. The ore production rate is pegged at 30,000 tonnes per day; that's 11 million tonnes of ore annually, producing approximately 155,000 tonnes of concentrate per year. Two phases are envisioned, utilizing conventional truck and shovel equipment, and mining methods that will incorporate drilling, blasting, loading, and haulage.

The treatment process at Morrison will likely involve a conventional crushing, grinding and flotation system. Ore will be processed through a conventional milling circuit consisting of a primary crusher, High Pressure Grinding Rollers (HPGR), ball mills and flotation circuit. Copper will be concentrated by flotation in large tank cells after grinding, then cleaned and filtered to achieve acceptable shipping moistures without thermal drying.

Pacific Booker Minerals recently had the Morrison ore tested at Polysius AG, in Germany to determine if there was an advantage to use HPGR versus a conventional SAG mill. The tests conducted at Polysius AG were to determine the size of the HPGR unit, power requirements and wear rate. The results indicate that the Morrison ore is very suitable for high pressure grinding with a significant reduction in power versus a SAG mill and very low wear rate. The reduction in power is approximately four MW and roller life expected to be plus 8000 hours. In addition the use of HGPR will also save approximately 10% in consumables. The decision was therefore made to use HPGR versus a SAG mill.

An ore delineation and pit geo-technical drill program has been completed at the pit site, along with 'Pit Geo-technical Investigations' and 'Pit Slope Design' reports. As well, a geo-technical and condemnation drill program has been carried out on the proposed waste management site and the process plant site.

Two alternative methods have been investigated to supply power to the project. Both originate at the existing B.C. Hydro Babine Substation located on the west side of Babine Lake in the vicinity of the Village of Granisle. From Babine Substation a new 138 kV transmission line could be constructed that would extend north along the western shore of Babine Lake, cross the west arm of the lake either overhead or via a submarine cable, then extend in a northerly direction to a new substation at the

Morrison Project site. The second, preferred option is to extend the existing line from the Bell Mine site. The 138 kV service, which was extended to the Bell Mine in 1971 by an underwater cable, is now energized at 25 kV and could possibly be re-energized to its design voltage. Pacific Booker is conducting an integrity test on the line, and if that is satisfactory, power will be available 22 kilometres away. If the line is not of sufficient quality, there is the option of drawing power from the Babine substation on the west of Babine Lake, about 45 kilometres away. The disadvantage there is not so much the cost of construction, but the additional environmental work required. If one runs a line adjacent to an existing road, the environmental assessment work is pretty straightforward; however if power has to be brought from the Babine substation, a little more environmental work is required.

The environmental baseline and Project impact assessment studies are well advanced. Studies commenced several years ago and in 2006 Pacific Booker Minerals retained Rescan Environmental Services Ltd. to consolidate prior studies and to complete the requirements for the Environmental Assessment Certificate. The environmental assessment will be used to apply for the mining permit for the construction, operation and maintenance, and decommission and reclamation of an open-pit mine at the Morrison project as per the Terms of Reference (TOR). Environmental studies include a range of fisheries, wildlife, aquatic, terrestrial, water quality, waste management, metal leaching and acid rock drainage, traditional use and traditional knowledge, archaeological and socio-cultural as well as the consultative process with the regulators, First Nations and local communities. These consultations are critical as Morrison is based within the traditional territory of the Lake Babine Nation, a band comprised of five communities: Burns Lake, Fort Babine, Tachet, Old Fort and Donald's Landing. Therefore, Pacific Booker is taking proactive steps to maintain a strong, positive relationship with the local First Nation communities in the Morrison area.

Development of the Environmental Assessment Certificate Application for a Permit to construct will occur in late 2007 following completion of the environmental fieldwork and impact assessment, concurrent with completion of the Project Feasibility Study. Permitting will be applied for concurrently so if all goes well construction will begin in mid 2008, with production starting late 2009, early 2010.

"We have a very good relationship with the First Nations," states Erik Tornquist. "Now, there are two components to a relationship with the

First Nations: one is the formal one, which is conducted through the Environmental Assessment office; the other is the informal one—that's the relationship building component. On that front, we have an excellent relationship. When we first met with the Lake Babine Nation, they asked us about the types of jobs we'll have in the mine," says Tornquist. The upcoming construction workforce is estimated to be in the realm of 450 people, with an operations workforce of roughly 220.

Tornquist's experience working internationally has helped him build a working relationship with the Lake Babine Nation. In Oman, Tornquist was mandated to have locals running the company within five years. So, he had to set up a training program. "For Morrison," he explains, "We created job descriptions and a training program for the mine, including the skills and the knowledge required. Then, we gave the Training Plan to the Lake Babine Nation. They have said that they appreciate it, because we're not coming in a month before we're going to start construction and asking for skilled labour. We're giving the Lake Babine Nation an opportunity to start looking at their human resources inventory and training requirements. Meetings have also been held with the B.C. provincial government to discuss training, because this government is very keen on assisting industry to train people because there is a shortage of labour in B.C. right now. As a result of the meetings we have provided the provincial government with the Training Plan. In addition Pacific Booker has submitted a proposal to the Federal Government for institutional training and an apprenticeship program for 18 Lake Babine Nation members. This proposal is part of the Aboriginal Skills and Employment Partnership program, managed by the Skills and Employment Branch of Human Resources and Development Canada.

The lifespan of an eventual mine (at the proposed production rate of 30,000 tonnes per day) is expected to be sixteen years. According to Tornquist, "the mine life could be longer, because we have 56 million tonnes, inferred, which is actually of a higher grade than the measured and indicated resource. So, that will come into play once the mine goes into production."

The current resource remains more than worthy of Pacific Booker's plans for a decade-and-a-half of mine operation. "The plant will produce the copper–gold concentrate, which will be shipped to smelters. There will be a molybdenum circuit and that molybdenum will probably be shipped to Endako for roasting," says Tornquist. A 26,500 tonnes per day open-pit molybdenum mine, mill and roaster, Endako is Canada's largest capacity molybdenum mine, and the only one producing molybdenum as its primary product. Endako is only 140 kilometres away from Morrison.

The diligent research, planning, and preparation that goes into every decision Pacific Booker makes is evidence of their commitment to actually operating a mine. We are confident that we're going to be able to raise at lease 50% of the money required for construction of the mine through equity financing, and the balance through the bank. At the moment, it looks like we're going to be able to start production on time."

During this important phase of mine development, the Morrison deposit is fortunate to have Pacific Booker's undivided attention. "We have other showings on some of our claims, but we have not done any work on them," Tornquist says. "The B.C. government Geoscience department has indicated a couple of showings on our properties. But, like I said, we have not followed up on that." Tornquist observes, "One of the problems in the business is that people are always saying 'I want to drill more, to prove up more ore.' But, you have to draw a line somewhere. You can drill forever and miss an opportunity of going into production. Then, once a mine goes into production, it's an ongoing task to plan your mining for every year. Then you can work on proving up more ore. Though, the first step is to go into production."

With the support of Pacific Booker's comprehensive management staff and directors, Erik Tornquist seems well-equipped to embark on this forthcoming, momentous step on behalf of the company. "Presently, our head office is not big. We have myself, John Plourde, a Technical Assistant, an Environmental Manager, and a Training Manager. John is a person that's obviously been making the money for us, so obviously he's of prime importance." Plourde has over 30 years of investor relations experience and has served as a Director of several public companies. Since December 1999, he has been a Director of Pacific Booker Minerals Inc.

William G. Deeks, PEng, Executive Director, Chairman of the Board, has 50 years experience in mining. From marketing, research & development, environmental management, health safety and accident prevention at work sites, approval of exploration budgets and work on project feasibilities, Deeks is also on the Governance committee of the Mining Association of Canada's "Toward Sustainable Mining" initiative in which Pacific Booker will participate. Most recently, he was Chairman of Charles Tennant Canada, a mining chemicals/collectors producer and chemicals distribution company. Gregory R. Anderson is Pacific Booker's Chief Executive Officer, President, and Executive Director. With a lifetime of business, corporate finance, investment, marketing, and brokerage experience, he has, since 1997, been instrumental in funding all aspects of Pacific Booker Minerals' growth as well as handling investor relations.

"Our Directors are important too," Tornquist explains. "Because they're all business people, and they run successful businesses, they're able to assist Pacific Booker to realize our full potential. Booker's Directors include Mark Gulbrandson, Executive Director, who is the owner and Chief Executive Officer of Apple Auto Group, a multi-million dollar business and a top 100 Ford dealer in the United States. Another Director is Dr. Dennis C. Simmons, DDS, who has a lifetime of experience in business, the medical profession, and a background in mining. And, William F. Webster is an Executive Director, who brings 40 years of experience in financial management, investment sales and corporate finance.

Overall, as a company with a market capitalization in the area of $100 million, Pacific Booker is strengthened by a formidable management and directorship working behind the scenes to develop and establish Morrison as a bountiful new copper–gold–molybdenum mine. Morrison's proven mineral resource, coupled with Erik Tornquist's precise and detailed plans for a realistic infrastructure, place Pacific Booker in a position as 'one to watch' in the mining sector.

QUADRA MINING LTD.

"It's evolved," says Morris Beattie, CEO of International Molybdenum PLC (InterMoly), 82% of which was recently acquired by base-metals developer and operator Quadra Mining Ltd. Despite the change of hands, InterMoly's former flagship property, the Malmbjerg molybdenum deposit in Greenland, is still under Beattie's able and talented supervision as he remains the head of the whole project for Quadra.

"InterMoly will still exist as an entity, but it is now a subsidiary of Quadra Mining," says Beattie.

InterMoly was formerly a publicly–trading company in the U.K., Beattie explains. "Quadra made an offer for all of the shares of InterMoly in March 2007. As of June 2007, when the offer closed, they held about 82% of the shares and 90% of the warrants; they are firmly in control, and are exercising their rights to acquire all the InterMoly warrants outstanding. As of June 2007 we have de-listed the company. So, InterMoly still exists as a U.K. company, but it is an unlisted company owned predominantly by Quadra."

Malmbjerg's history with InterMoly began back in 2004, when a British-based international mining development company, called Galahad Gold PLC, acquired an exclusive license from the Government of Greenland to explore and develop two molybdenum projects in eastern Greenland—the Malmbjerg deposit, and Flammefjeld (about 500 kilometres south of Malmbjerg). Hence, Galahad established International Molybdenum PLC as a special purpose subsidiary to concentrate on the development of the deposit at Malmbjerg.

Acquiring the exploration licenses for both the Malmbjerg and Flammefjeld projects from Galahad, InterMoly became a public company in April 2005 and its shares were admitted to trade on the AIM (Alternative Investment Market) of the London Stock Exchange in July 2005. "Essentially, Galahad was the parent company of all of this. They issued shares to take it public, but still owned just less than 80% of InterMoly. So, that's how the deal with Quadra eventually got done," states InterMoly's Beattie. "An irrevocable undertaking to accept Quadra's offer was received from Galahad resulting in the immediate acquisition of approximately 78% of the InterMoly shares and 25% of the InterMoly warrants."

A November 2005 NI 43-101 compliant technical report for the project estimated a measured and indicated resource of 560 million pounds of contained molybdenum. As an advanced base metal project, therefore, the promising Malmbjerg seemed a good match alongside the rest of Quadra's impressive property portfolio.

Formed in 2002, Quadra Mining Ltd. (TSX: QUA) is a British Columbia corporation based in Vancouver, known as a copper–producing company whose principal asset is the Robinson Mine in Ely, Nevada. From the beginning Quadra had a strategic plan of becoming a mid-tier base metals development and operating company with interests in production, development and advanced exploration copper properties. With a business model of growth by acquisition of non-core assets from larger producers, accessing development projects held by exploration–focused companies and seeking merger opportunities within the small to mid-tier sector – things certainly seem on track to meet their goals. Especially considering their latest maneuvers resulting in the acquisition of InterMoly and its potential molybdenum producing giant, Malmbjerg.

And the acquisition of Malmbjerg is one of several that Quadra has closed since its initial 2004 acquisition of the Robinson Mine. The company raised CDN$145 million – one of the largest financings in the resource sector that year, to acquire the Robinson Mine for a consideration of US$14.2 million. In 2005, following a successful restart of the mine, Robinson produced 145 million pounds of copper and 81,000 ounces of gold. Subsequently, the strategic growth plan continued with the purchase of the development-ready Carlota Project in Arizona, which is now under construction and will produce copper cathode starting in 2008, and the advancement of the Sierra Gorda copper–molybdenum project in Chile towards the pre-feasibility stage.

Quadra has set a mandate to triple its current annual copper equivalent production to approximately 500 million pounds to achieve the critical mass of a mid-tier producer. Add to that potent mix the prospects at Malmbjerg, and you have a company that is striving to position itself as a solid investment vehicle for movement within the base metals industry.

"As you see, it's definitely evolved," says Beattie "InterMoly will still exist, like I said, as a private company and an 82% subsidiary of Quadra. And I'll still be the head of that subsidiary. I gave that undertaking to them when they were doing the acquisition, rather than sort of just saying, 'Okay, that was fun, and now it's yours.' I agreed that I would stick with it and advance it. At least for a year, to make sure that there's continuity on the activities.

"So, right now, we have moved our offices into the Quadra headquarters. We are all here as part of this new and very exciting team."

Now, working as one big, expanded family, Beattie is understandably pleased to be staying on with a moly project of Malmbjerg's calibre. "That's right," he says. "When we looked at this project—as Galahad, at that time—we were very bullish on molybdenum. We thought it was a good commodity to get involved in. And, you know, obviously we made a good call there.

"As Galahad, we were shareholders in Northern Dynasty's Pebble project, a large gold–copper deposit, up in Alaska. Though Galahad recently sold its interest in the Pebble project, we also have an interest in a uranium company, Uramin Inc., which has just announced it is being sold, too. But, what Galahad did as a deal for InterMoly, it accepted Quadra shares as payment— so it's now a shareholder in Quadra and in Uramin. Galahad was in the mine investment/mine development business and we'd looked at other projects in Greenland in the past. Because of that previous experience, we had a one-step advantage on a lot of other companies; in terms of knowing how to operate there."

With InterMoly created for the express purpose of developing Malmbjerg, the company did exactly what it was formed to do, notes Beattie. "It's done it perfectly. Acquired the project at the start, developed it far enough to show that it's a very real resource, that it's a project that has development possibilities, and then found a company, Quadra, that has the wherewithal to advance the project from here to production."

Located at the Werner Bjerge Mountain Range, on the east coast of the island nation of Greenland, approximately 600 kilometres northwest of Reykjavik, Iceland, the Malmbjerg molybdenum project boasts 'Climax-type mineralization,' which was first discovered in 1954. Over the period from 1955 to 1979, the property was explored by means of 147 diamond drill holes totaling 22,877 metres, and three underground adits totaling 1329 metres.

Later, in 2005, according to independent geological consultants Roscoe Postle Associates, Malmbjerg was reported to contain 560 million pounds molybdenum, indicated at a cut-off grade of 0.12% MoS2.

The results broke down as follows: measured and indicated resources of 217 million tonnes at a grade of 0.20 % MoS2 with an additional inferred resource of 12 million tonnes at a grade of 0.15 % MoS2 , using a 0.12% cut-off. This estimate includes a higher grade measured and indicated resource of 33.8 million tonnes at a grade of 0.28% MoS2, using a 0.25% cut-off. This resource estimate, compliant to the Canadian NI 43-101 regulations, incorporated historical data as well as new data from the 2005 program.

"While it's a large resource; the quality it really has is grade," says Beattie fervently. "It has a considerably higher grade than most large moly projects under development. Of course the costs of mining in Greenland will be higher than in other places, such as Australia and Nevada. But, be that as it may, it's not going to be double the costs of working in those other countries, while our grade is roughly double what the other grades are out there."

Moreover, in 2006, InterMoly initiated a pre-feasibility study on the project. A report from independent mining consultants GR Technical confirmed the technical viability of developing Malmbjerg as an open-pit mine. A plan of operation, by InterMoly, estimated 15,000 to 20,000 tonnes of molybdenite concentrate to be produced per annum. Thus, an open-pit operation is expected to have a 15 to 20 year mine-life, with production expected to commence in 2011.

"As we speak, there are Hercules planes flying from Iceland to Greenland, taking in our drill equipment, our mining equipment and fuel," says Beattie. "They do, maybe, two or three flights a day for a period of a week." Momentarily, Beattie is also heading out to the property on a flight himself to meet up with the equipment for "our 'kick off' site-visit for the year. And, I'm taking over some Quadra people so that they can become more familiar with all aspects of the project."

This buzz and excitement surrounding Malmbjerg is proof enough that times have changed. As Beattie points out, "For a number of years, you could never have generated any funding for a molybdenum project—when it was down around $5 or $6 a pound and there was an adequate supply. It was just all of a sudden that the supply in no way matched the demand. Only then did it seem like a good commodity to get into. But, you know, there have been many companies who, later on, if you like, 'jumped' onto the moly bandwagon. We were already well into it by then."

A PhD graduate from the University of British Columbia, Beattie explains, "In between my studies BASc, MASc through to PhD, I was working as well. I've always been interested in mining, predominantly on the metallurgical side. And, interestingly enough, the very first job that I had in an actual operating mine was at B.C. Molybdenum. So, it's like my career has come full circle, 35 years later."

InterMoly's Beattie is not only an early believer in moly, but could also be described as a trailblazer in Greenland's mining scene. "There are a few companies working there," he says. "There's a lot of oil exploration and at the moment, there's exploration ongoing for diamonds and rubies, and lead and zinc. Previously, there was an operating lead–zinc mine on the west coast, called the Black Angel. We had another project in Greenland prior to this one, so we were familiar with Greenland, with the regulators and how to work a project there."

Certainly, some might find a lack of discernible infrastructure concerning, but as Beattie says, Greenland isn't unknown territory for his company. Nevertheless, in the case of Malmbjerg, is there infrastructure? Beattie chuckles as he says, "No, it's absolutely in the middle of nowhere. We bring in generators for electricity and so on. When this ultimately becomes an operating mine, it will be diesel-generated power and everything we need for the operation will come in by ship during the summer."

Additionally, he says, "Within Greenland, there are no cities to speak of anywhere near us. The total population in Greenland is around 60,000 people. Fifteen thousand of those live in the capital which is Nuuk, over on the west coast, the opposite side of the country from Malmbjerg. Whereas, on the east coast there are two villages – the nearest of which to us is Scoresby Sund, and there are only a few hundred people living there."

Furthermore, due to the lack of nearby infrastructure and local manpower, "Most people are brought in from other areas for the simple reason that there aren't very many people living here. We've hired as many residents as we could on previous drill programs. We had twelve locals working with us over the course of the summer and that probably represented most of the available employable population of Scoresby Sund. So, you hire—when you can. " Otherwise, "We fly personnel in from Iceland and most of our people, for instance our diamond drillers, are Canadian. Greenland is a self-governed Danish territory and there's a place called MestersVig which is a Danish navy and army base, on the east coast. There's an airstrip there and some infrastructure and so on. We fly in and out of there by helicopter. We commute to the project site."

Compared to Canadian properties, working in Greenland has several distinct differences to note. For instance, there are no Aboriginal claims—"No, nothing like that in Greenland at all." Permitting is simplified, too.

"They have regulations, as respectful of the environment as Canada, but the processing is more straightforward," says Beattie. "They have one Bureau of Minerals and Petroleum so, whether I need to apply for a drill permit, a blasting permit, or ultimately for an exploitation license, I go and talk to the same people and they take it from there. I have found them to be very cooperative and helpful," he says assuredly.

One of the similarities Greenland has with Arctic Canada is its climate. "Our peak season is summer—July and August. Normally, at this time of year, we would expect night time temperatures of say -5 °C and daytime temperatures of +10 °C. But, currently, we've got an unusually warm period. Now, having said that, we were there drilling in 2005 and it was between +16 °C and +18 °C a lot of the time, while we were out on the drills. So, you know, when it's summer, it's really quite pleasant."

With these amenable conditions and regulatory climate positives, in addition to the staggering moly prospects in the ground at Malmbjerg, Morris Beattie confirms, "In a nutshell, our focus is on getting this project approved for production. Effectively, that's it."

Quadra's core asset is its operating Robinson Mine - an open-pit copper and gold mine located in eastern Nevada, approximately 11 kilometres west of the town of Ely. Beginning in 1994, Magma Copper, which was later taken over by BHP Copper Inc., constructed and started up a modern mining and sulfide concentrating facility at a reported cost of US$480 million. The mine was closed in 1999 because of low metal prices. Quadra acquired it in April 2004, and restarted production later that same year – on budget and ahead of schedule.

Currently, the Robinson Mine has a mineral reserve of 122 million tonnes at a grade of 0.69% Cu and 0.26 g/t Au. Estimated annual production exceeds 125 million pounds of copper and 60,000 ounces of gold, and the mine has an estimated nine-year lifespan.

Since 2005, its first full year of commercial production under Quadra ownership, Robinson has been generating operating cash flow, and as of June 30, 2007 the mine has produced 340 million pounds of copper, 223,000 ounces of gold and 260,000 pounds of molybdenum.

Quadra is continuously looking for ways to improve production and to become as efficient as possible at the Robinson mine. Since its acquisition the company has made significant upgrades to the mill by adding a gold recovery circuit, a molybdenum recovery circuit and a new product drying system.

Quadra is also looking for additional reserves to extend the projected mine life at Robinson. The company has begun by exploring at the Veteran Pit, plans to move to the Ruth pit, and is test- drilling the massive tailings left from previous decades of mining.

Other assets under development with Quadra are the Carlota and Sierra Gorda projects. The Carlota project is a permitted, bonded and financed late-stage copper property currently in construction. It is located in the Globe–Miami Mining District of Arizona—one of the most prolific copper mining regions in the world. The project is located among several historic mines and is contiguous to the BHP Pinto Valley Mine.

Carlota has probable mineral reserves of approximately 86 million tonnes grading 0.45% total copper. The project would comprise a run-of-mine, heap leach project, producing cathode copper by solvent extraction/ electrowinning (SX/EW)—a two-stage process that first extracts and upgrades copper ions from low-grade leach solutions into a concentrated electrolyte, and then deposits pure copper onto cathodes using an electrolytic procedure.

In due course, Carlota is looking at approximately a nine year mine life, plus two years of residual leach. Its plant is being designed to produce up to 75 million pounds per year, with initial expectations of copper production amounting to 28 million pounds in 2008, ramping up to 75 million pounds per annum in 2009.

The Sierra Gorda project is comprised of eight adjacent properties that cover an area of approximately 23 square kilometres in a region of northern Chile that has hosted a number of other significant copper mines and resources including BHP Billiton's Spence project, 10 kilometres to the northeast; Antofagasta Holdings' El Tesoro mine, approximately 15 kilometres to the southeast; and the Lomas Bayas and Mantos Blancos mines to the southwest.

An NI 43-101 compliant report for the project estimated an indicated and inferred resource of approximately 5.0 billion pounds of contained copper. Plus, recent deep sulphide drilling returned significant intervals of copper, gold and molybdenum mineralization. The length of the intervals and the high grade supports the idea that there is a substantial new zone of classic porphyry copper mineralization.

Of course, mine development at the Malmbjerg moly project in Greenland is also on Quadra's plate. "Right now, we're working on an interim feasibility study that we'll target for completion by early 2008. The definitive feasibility study would be targeted for completion by the end of

2008," says Beattie. "Construction would take a couple of years, so it would be early 2011 before the mine is ready to operate. That's our target timeline."

Beattie continues, "We've started a couple of feasibility, pre-feasibility type studies before, and each time we found things that improved the project significantly. Hence, the first thing we're doing is some comparative trade-off studies with a few alternatives. Our first step was defining the resource, which we did in 2005 with an extensive drill program. Following the resource definition, you go in and determine the best way to complete this project, in terms of how you develop the mine, the roads, how you transport ore from one place to another. That's what we've been going through."

During some of the original pre-feasibility studies, InterMoly admittedly ran into some complications. "We started off using historical information and it lead us to looking at this as an underground mine development, with access from the north," says Beattie. A 15,000-ton-per- day, underground operation was envisaged with the projected annual output representing around 4% of total world consumption of molybdenum. "Yet, as we did the resource drilling, we came to the realization that an underground operation was not the best way to develop the project. We realized, in fact, that the deposit lent itself very well to an open-pit development because of its configuration, with access from the south—making it a much lower risk operation."

In March 2006, InterMoly announced that it was evaluating a revised development plan after a deferral of the original feasibility study for the underground system. The open-pit method was seen as a lower cost way of developing this deposit, eliminating the need for a 10-kilometre underground tunnel. And, it is believed, that opting for an open-pit operation at Malmbjerg will provide for larger-scale production, better recovery of the resource, greater grade control and will reduce capital and operating costs. With so much to gain, the company began undertaking a new pre-feasibility study.

"That was a tough time though," say Beattie, "to convince the investment community that stopping one study and starting another that went in a different direction – was a good thing to do, that it was in the best interest of the company. It took a while for that to sink in."

At the moment, Morris Beattie and his colleagues seem to be closing in on a potentially happy ending for Quadra's newest project, making Malmbjerg a tangible success story for InterMoly and himself. "It has been very satisfying personally, because many of my peers have congratulated me on having taken InterMoly and Malmbjerg through these steps," notes the 59-year-old.

Now this budding molybdenum success can be further developed by Beattie, under the umbrella of Quadra, and naturally with the considerable exerptise of Quadra's skilled management team who have hands-on operating, development and financing experience in nine mine start-ups.

Morris Beattie heads this team—once concentrating on doing what was best for InterMoly's shareholders, now he's deliberately focused on doing the same for Quadra's shareholders as well. "You know, from my role, they're one and the same. If we do a good job, it becomes more valuable and, everyone benefits from it.

"Although I've spent my whole career in mining, I'm not interested in operating a mine. I like taking things from discovery to development—here's a bunch of rock and it seems to have some value, to figuring out how you recover it, and how this all comes together and advancing it. Sort of exactly what's happened with Malmbjerg. That's what I like to do."

A driven and resourceful individual, he is an ideal guardian to oversee Malmbjerg's continuing progress. He admits, "My pleasure is in making things happen; in advancing the development.

With Beattie's burning drive for advancement and 'making things happen,' combined with Quadra's well-known operating and financing expertise, the Malmbjerg molybdenum deposit appears to be in very good hands.

ROCA MINES INC.

Scott E. Broughton, the President and CEO of Roca Mines Inc., is intent on maximizing the vast potential of their MAX Molybdenum property. Already permitted and aiming for initial molybdenum concentrate production in 2007, MAX Molybdenum is Roca's primary project: a 50 million ton primary moly deposit, which Broughton ponders could be home to more—much more. Could it be an upper offshoot of something much bigger, lurking beneath Trout Mountain, British Columbia?

"Absolutely, there's a very, very distinct possibility," says Broughton. "And more than that, it's an extremely compelling exploration project, given that we are an exploration company as well. We have looked at this deposit and said to ourselves, 'there's a concentration of metal in this deposit which is outstanding,' that really speaks to the grade of some of our high grade zones. The intensity of stock work and the repeated phases of mineralization in this deposit all point to a deeper source. And then, in the same context, MAX is a porphyry intrusion related deposit, and we think that this all points to something at depth."

Knowledgeable, thoughtful and direct, Scott Broughton certainly presents himself and his company well to investors. With a property like MAX to talk about, he is self-assured that Roca is on to something "big."

Identified as a 'fast-track mining opportunity,' Roca seized the chance to develop the MAX Molybdenum mine. They have advanced the project dramatically since completing its acquisition in late 2004, and Roca's construction work is continuing toward production.

A professional engineer by trade, with over 20 years in the mining industry, Broughton's career notably spans exploration; mine design, development and operations; consulting engineering and project management; specialist geotechnical services; business development; and the formation and management of public companies. He has a Bachelor of Science degree in Mining Engineering from Queens University, Ontario, and is a member of the Association of Professional Engineers and Geoscientists of British Columbia.

"I've had a life-long interest in rocks and geology, and my father was an engineer. We were nomads; we lived in Winnipeg, Toronto, and then went to Africa for some of my teenage years. My father worked on industrial development projects in Zambia. But there's a big copper belt in Zambia and I had lots of exposure to geology and mining—it was great."

Therefore, Broughton explains, "mining and engineering were a natural thing for me to pursue. During summers off university I worked for mining contractors developing mines and deep shafts in Ontario, Alberta and NWT. I graduated and worked primarily in engineering consulting, with a lot of that work for senior mining companies around the world. I got lots of exposure there. "Later in life, I moved on to finding out that there were opportunities other than just purely technical. One of the skills that I'd developed as an engineer was putting project teams together, and finding technical people. I just sort of gravitated towards 'talking the talk' and finding investors to help fund these projects."

From an interest in rocks to handling the business side of mining, Roca's Scott Broughton has seen it all. "it was kind of an interesting evolution. I became involved in public companies, going back through my career, at least 15 years ago," he says. "I took my first company public only weeks before the massive Bre-X fraud that killed the junior mining business for quite some time."

Bre-X—a junior Canadian mining company based in Calgary—perpetrated an elaborate scam when it reported that it had bought an enormous gold deposit in Indonesia. Originally a penny stock, Bre-X's stock price soared on the TSX to a high of $285 (CDN), eventually reaching a total capitalization of $6 billion. But independent testing eventually proved the gold results to be bogus and Bre-X collapsed in 1997, making the shares worthless and bringing the entire mining and exploration business down with it.

"My first company was an exploration company" nods Broughton soberly. "And Bre-X killed everybody's interest in mining investment and then, of course, the price of gold dropped dramatically after that; that was my MBA in timing, business, finance and investment—having to live through that downdraft, if you will."

Indeed Broughton survived the 'downdraft' and now the 45-year-old runs Roca, a thriving company, primarily focused on British Columbia-based properties. Even though it's tempting to consider sites elsewhere around the world, including Africa, Broughton concludes "I won't ever say never, but we decided early on that British Columbia has lots to offer and it's in our own back yard. It's easier to manage than some of those far away places. There are world class mines here, but because of some of the history in this province it's been under-explored to a certain extent."

Moly, however, wasn't always at the forefront of Scott Broughton's mind. "We're basically metal miners. We're not that interested in coal or resources associated with softer rocks. We're into metals—base metals and precious metals, and now noble metals like molybdenum. You know, we've been in that exploration mode for those metals in British Columbia since Roca was founded in 2002." The company was started by Broughton and a business partner, John Mirko.

Mirko has 30 years experience within the resource and mining sectors. "John is a prospector," notes Broughton, "He doesn't have a geology degree. He's self taught, a self-made man. He's been very, very good at field work and the logistics of organizing field programs, how to run a camp, how to get helicopter support, all the things that have huge costs for exploration companies. More than that, he's very smart on what makes a project potentially work—because it's going to need some infrastructure to support it. You can't just go and find some small mine in some remote place and hope somebody will build a road for you. That doesn't necessarily work. You have to be mindful of that. John's been very good in terms of making sure that the projects that we tackle have at least that kind of potential. Otherwise there's no point exploring them. He's had a hit list of targets known about in the province, and MAX was one of them. He identified it as an opportunity.

"We found the prospectors that had acquired the key claims there, then John and I tackled bigger issues of negotiating the right kind of option agreements, and we've been successful together making these approaches to bigger companies and getting the things that we need. That sums up John's skill set—he identifies compelling target projects and once acquired spends a lot of time in the field making sure things work the way they should for this company."

Along with Broughton and Mirko, Roca later attracted David Skerlec. "The three of us work as partners," Broughton explains. Skerlec, MBA, is the Chief Financial Officer with over seven years experience at a Vancouver-based investment dealer, most recently as Vice President, Corporate Finance where he specialized in exploration finance. His diverse background includes over five years experience in the junior mining business where he supervised exploration and production at a placer gold mine.

"Dave, John and I run this operation like a partnership, even though it's a public company. Dave comes from the corporate finance world, but his family was involved in placer gold mining in the Yukon. He understands the exploration business, the drilling business, the logistics we have to go through, and the mine building issues. So, he's a great C.F.O. to have and a great link to the finance world."

In addition, John Kiernan, (P. Eng.), Roca's Vice President, Mining, brings to the table 22 years of operating and engineering experience in the mining industry. He was a principal mining engineer for eight years with AMEC/MRDI, and spent six years as a mine planning engineer and project manager with INCO in Sudbury, Ontario and Thompson, Manitoba.

"John came to us a year ago; he and I were at school together studying mining engineering. I've got a lot of confidence in what he can achieve. He spent most of his career in a pure engineering environment doing feasibility studies, designing and organizing mines, as well as 'hands on' early stuff with bigger mines. He has a good understanding of how a real mine has to work. Together we have worked through a lot of development issues and done a lot of planning for how MAX will work initially and how it could evolve. He gets a lot of support from the site, most of our contractors and employees are based there obviously, while we go through these development phases, says Broughton."

With an impressive team, Roca's Broughton notes that back when the company was founded, "we thought that the lull in the industry and investment would soon break, and one day we'd be there enjoying the kind of metal prices we have today—which is really an indication that there haven't been any new mines built for a long time and yet the world continues to grow. And that's why metal prices are up, including molybdenum."

Broughton smiles, "If I could sum it up—I had extraordinarily bad timing with the first company that I took public, but I think we're well positioned with good timing this time around." Especially if you take into consideration that MAX has a global measured and indicated resource of 42,940,000 tonnes, grading 0.20% MoS2 at a 0.10% MoS2 cutoff; boasts in excess of $3 billion contained metal value (at $30/lb molybdenum, not including any inferred resources); and remains open at depth.

Back in 2003, with molybdenum prices on the rise, Roca began a series of negotiations which led to the acquisition and consolidation of a 100% interest in the molybdenite deposit known as 'Trout Lake', located 60 kilometres southeast of Revelstoke, British Columbia. "There were a bunch of problems with the name, notwithstanding there's a number of projects in Canada that are called Trout Lake. That was the fundamental reason we needed to change it," says Broughton of the renaming of the deposit to

'MAX Molybdenum.' "I think it's a good choice of a name. There are some parallels there to the history of Climax. But, basically, it's just a name that struck us as suiting a maximum opportunity for Roca Mines to move ahead quickly."

Distinguished by its significant high-grade resources within a much larger deposit, MAX has been the subject of extensive exploration and engineering programs over the decades, conducted by Denver, Colorado-based Newmont Mines Ltd. and Esso Minerals Canada Ltd. in the late 1970s and early 1980s. Despite the extent and quality of the previous work, the project was never put in production by the joint venture.

Nevertheless, according to Broughton, the price of molybdenum wasn't the prevailing factor in Roca's decision to ultimately produce MAX themselves. "This is one of the cornerstones of our story when we tell it, and I have told it about 500 times, what we knew about the project was that (and here kudos go to John Mirko) it wasn't just a well-defined deposit—meaning that the previous operators, Newmont and Esso, had done a lot of work there drilling it and proving it up—but, anecdotally, it also has some very high grade zones in it, and that makes it a kind of unique deposit among the molybdenum deposits we know in the world today.

"That high grade, and the fact that Newmont had excavated this large production-sized adit, or tunnel, right in the heart of the deposit, provided really the two leading facts that we knew we needed to build on to advance a property and make a mine," says the Roca President. "So, one was that key piece of infrastructure to get into it, and the other was that the grade was probably going to be high enough for us to support a small-scale mine operation there. And all of that was independent of the price of molybdenum.

"At the time we acquired MAX, moly was around $8 a pound and it had been edging up from the $4, $5, $6 a pound range. It didn't light anybody on fire, you know. Yet, we had no special knowledge here—it was just our observation that the deposit had some of the key attributes that we felt would put it in good stead for building, initially, a small mine. "That led us to striking a deal; a classic option arrangement where we had the right to explore the property, and ultimately own it if we lived up to our end of the agreement. Early in 2004, we realized that we had to consolidate the ground and one way to do that was also to negotiate with Newmont to acquire the remaining property they had in the area, and more importantly, all the data that they had for this project. Because they were the only ones that held all of that valuable information, back in Denver.

"This comprehensive database included all the geological logs, all the assay information, all the engineering they did, which was substantial. Plus the environmental work that they did was also substantial. All of that was extremely valuable, much more valuable than just the $15 million that they had spent on the site. We bought it in another agreement, another option with Newmont, and paid them only a small amount of money for it—again, moly was at $8 a pound and they were essentially disinterested in it," explains Broughton. "That one step really gave us a leg up to do lots of detailed engineering, and start our own environmental work, our own baseline work, and in part it was one of the ways we achieved the fairly rapid permitting of the property—because there was 25 years of solid background information that started with Newmont."

Newmont's data also pointed to the possibility of MAX's deposit being only the beginning of something bigger. "Absolutely—if you review the exploration history and the drill logs that we acquired from Newmont, there are three holes that just kind of manage to stray below the known resource and they continue in mineralization. It's not spectacular, like the kind of stuff we're mining initially, but it's evidence that there is some more at depth there," Broughton says. "We decided to try to review some of the geological models that are out there. One of the parallel models that kept on coming up was a comparison between MAX and a deposit called the URAD, which is in Colorado; again it's a primary molybdenum deposit."

The URAD deposit is the uppermost deposit associated with the famous 700-million-ton Henderson deposit, owned and operated by Climax Molybdenum. Henderson has operated since 1976 as a primary molybdenum mine, producing over 37 million pounds of molybdenum in 2006.

Climax started mining the relatively small (approximately 13.7 million tonnes of 0.35% MoS2) URAD ore body, while geologists theorized that the deposit was an offshoot of a larger and deeper deposit. After several years of exploration, a 1965 deep drill hole tagged what is now the Henderson deposit within Red Mountain. Forty years later, the Henderson mine is operating at over 30,000 tpd (tonnes per day), producing nearly 10% of the world's molybdenum.

"Henderson's still in operation. It's one of the world's largest underground mines and one of the world's largest molybdenum producers, single source. It made Climax, and then Phelps Dodge, into the force that it is today in the molybdenum business," notes Broughton. "So, that geological thesis could bode very well for us. And it's not just specifically Henderson; it goes to how porphyry systems work in general, and we see many similarities to that modeling in other kinds of porphyry deposits—copper and gold deposits, for example."

Of course, Broughton isn't just banking on theoretical comparisons to drive the speedy production of MAX. "We needed to attach some credibility to that story and we did that by assembling a senior exploration advisory board at Roca; a group of guys that are all senior. In this business, 'senior' means men in their 70s and 80s, because you know those guys were the active geological explorationists back in the 1960s and 1970s, and nobody's gone looking for big porphyry deposits since then, right? So, we hired guys that came from Climax and Henderson and other people that were extremely knowledgeable about those kinds of systems. They're the ones that are really guiding the first foray here, with the drill holes that are going to be drilled this year."

The advisory board consists of John Baker, a senior geological advisor to Roca since 2003, who also managed growth of JT Thomas Drilling Ltd., a successful international drilling company; Dr. N. C. Carter, P.Eng. who has extensive experience with B.C. moly deposits, such as Endako and Trout Lake/MAX and has reviewed Climax and Henderson; Don Davidson, M. A.Sc., based at Climax in Colorado, who was resident geologist at the major underground exploration project at Yorke Hardy, now the Davidson deposit; David Jonson, MS Geol., who held the position of Chief Geologist at Climax and explored the URAD-Henderson deposits; and Terry Macauley, P.Eng., who worked with Newmont, and was responsible for the exploration and evaluation program at Trout Lake.

Meanwhile, Roca retained Hatch Associates Ltd. to conduct independent engineering studies and preliminary economic assessments for a 500 tpd operation at MAX. The resulting cost estimates and financial models were further refined by Roca and its consultants, and were eventually filed as an application for a British Columbia 'Small Mines Permit' in the name of Roca's wholly-owned operating subsidiary, FortyTwo Metals Inc.

In November 2005, FortyTwo Metals was granted a production permit from the provincial Ministry of Energy, Mines and Petroleum Resources. Initial production will be focused on the rich 'HG Zone' in the centre of the area known as 'B-Zone' and will produce a readily saleable premium specification concentrate of approximately 95% MoS2. FortyTwo Metals purchased a 1,000-tpd mill and concentrator, along with other related equipment, for the MAX Molybdenum project. The mill, which was located approximately 380 highway kilometres south of MAX, was dismantled and moved in pieces to the MAX site. The mill acquisition included buildings; crushing, grinding, and flotation circuits; plus complete engineering drawings to allow for rapid reconstruction.

Roca's initial phase of mining is expected to produce approximately 1.5 million pounds of contained molybdenum from each production run of 72,000 tonnes. In 2007, Roca intends to complete back-to-back production runs resulting in the production of approximately 3.0 million pounds of contained molybdenum. Using a campaigned mining-milling approach, the company plans to recover much of the 260,000 tonnes of 1.95% MoS2 contained in the HG Zone within the first few years of production, and then make expansion decisions based on molybdenum prices during 2007 and beyond.

Roca is also advancing successive phases of mine development, allowing for rapid expansion of the operation using internally generated cash-flow. Milestones met so far include underground mine development, installation of electrical systems, construction of a tailings facility, mill upgrades, and other infrastructure improvements. For Phase II, development of a second adit to the deposit is currently underway.

Future exploration will focus on increasing MAX's resource, both at depth and in areas surrounding the main deposit. Exploration will be guided by Roca's aforementioned advisory board of porphyry molybdenum experts. However, Roca's accelerated exploration program is not expected to impact mine development or planned production. With the Phase I mine permitted, all systems are literally 'go.'

"Permitting was arduous," notes the CEO of Roca Mines. "But, you have to make an investment of time and money to put together a comprehensive application. And, the problem with going to a greenfield site, for example, is that you have to start with a minimum of a year's baseline data and the more historic data that the regulators have to review, the better." In the case of Roca's MAX Molybdenum project, "We could show them 25 years worth of data, so that was great. And then the second attribute is the location of this project because, with respect to permitting, this mine is not an unusual thing in this area. We had done a lot of community consultation and we had great support for the project. So, it all basically worked out perfectly."

With regards to community consultation, this includes both Aboriginal and non-Aboriginal peoples. "We've done most of our public presentations at the Trout Lake community, only a few kilometres away from our site. The First Nations issues in British Columbia are not to be underestimated. Most of the province is still in a process, and still in land claims, and subject to overlapping land claims as well. We actually lie within the traditional territory of the Ktunaxa First Nation, who are based in Cranbrook." Admittedly, Broughton explains, the Ktunaxa are "a long way away from us, but they're in an area that is well aware of the impacts of mining and the benefits of mining. They reviewed our application and made comment to it, and we responded to those comments and did additional work. We submitted archeological studies to provide some comfort for them, and to confirm that there wasn't something that was a missed opportunity at this site."

Permitted and ready for the next steps, Roca's Broughton says, "I think the reality has set in now that many of the projects that are around, if they're not in permitting, then they have almost an indeterminate length of time to get there. And we just really focused on our opportunity which was to get something going as quickly as possible. Our next problem after that was to convince investors that a small mine was a good thing. People who bought into the moly price and cycle often gravitate to bigger is better. We've been challenged by that a little when we want to attract attention to us."

Regardless of perceivable mine-size, with a name like MAX, it can be tricky to imagine a company having any trouble attracting maximum interest, especially with such a seemingly generous deposit of moly onsite. "The bottom line is that, yes, this is a small mine initially. It's certainly our first phase; our first year of mining in 2007 is going to deliver three million pounds of molybdenum—from a deposit that hosts 113 million pounds in the known resource," says Broughton.

Even with this known resource, the potential lifespan of MAX, Broughton points out, still depends on metal prices. "If you asked us six months ago, I guess our thought was that this moly price is only going to be a window of opportunity for maybe two or three years, because of all these other projects and how they're being talked about. But, if you ask us now, I think our opinion—just having come back from a molybdenum conference in New York—is that our window of opportunity is actually more like four or five years, because it's quite clear that most of those projects won't get financed right away, won't get through the permitting process right away, and then they have to build at a time when the rest of the world is busy with construction.

"So, our mine plan now is: let's expand and make our mine bigger, faster. We're targeting what we think is real—the potential to produce about five million pounds in 2008, and probably sustain that for a four or five year period. Of course, we've been asked to increase it to a minimum of a ten year mine-life. Most investors like to hear that. But, that's juxtaposed against this outrageous concern about what the moly price is going to be after four or five years. We basically just drew a line in the sand as, even after that stage of mining, we still only would have recovered about 27 million pounds out of the existing 113 million pound resource. Everybody could imagine that there's future expansion and future opportunities to exploit more of that deposit, depending on the molybdenum price, obviously."

Broughton continues, impassioned, "I believe that there is a 10 to 15 year mine-life at this site. We might have numerous expansion phases—we're already sort of between Phase I and II at the moment. And I could see us easily doing a Phase III and a Phase IV. It seems as long as metal prices are favourable, the MAX mine has the potential depth to deliver a pronounced moly supply—placing Roca Mines in a fortunate position. Yes, I think we're extremely lucky," says Broughton, "with all these attributes we've covered, including the good fortune that molybdenum prices have gone up at this time; all of that's worked out very well for us. But, you know, in this business you really have to make your own luck. You have to be positioned to be able to take advantage of those things."

Broughton confidently anticipates that Roca's production goals are on track to make the company the first new primary molybdenum producer in Canada. "If you look at the horizon of molybdenum projects that are talked about right now, there are very few that are talking about tomorrow's production. And there's even fewer that are also expanding on that and saying, 'Look, there's a tremendous amount of blue sky attached to this, too.' Most of them are already big and low-grade deposits that haven't got much else to offer in terms of adding to the size of the resource."

Broughton considers Roca to be the exception. He surmises, "We have this mine we're developing, we'll be the first new producers in the world of this valuable metal; it's pretty momentous to achieve that on any scale. The known resource is a smallish thing, but we believe it's monstrous underneath. This is the pragmatic and best way to build a mining operation, mine, and mining house. We've got the makings of a company builder. However, the whole thing may be for sale at the right offer. We're not egotists that 'this will forever be ours because we started it.' If it does get sold, we intend to leverage off that success and move on to other things."

Roca's inventory of properties isn't limited to MAX; the company's original flagship property is called Foremore. "It goes back to the base metals, precious metals aspect of the company. Foremore is the project that we actually listed Roca with—our lead property," says Broughton.

Foremore is located in one of the most active mining and exploration areas in British Columbia—the Eskay/Iskut Camp. This camp includes Barrick Gold's legendary Eskay Creek Mine, Cominco's historic Snip Mine and Novagold's Galore Creek project. Eskay is Canada's richest gold-silver producer with annual production of 366,000 oz. gold, and 16,000,000 oz. silver from 700 tonnes of ore per day.

As Roca's 'listing property,' it has been a staple focus of the company's exploration efforts dating back to the summer of 2002. The property was also the focus of exploration by Cominco Ltd. between 1989 and 1996, who were drawn to the area because of its abundance of mineralized boulders. But, Cominco failed to locate the source for the thousands of well-mineralized boulders.

In contrast, Roca's crews managed to discover numerous showings and mineralized zones in outcrop. "It's extremely exciting from an exploration point of view, in terms of what it's already delivered," says Broughton. "Our problem has always been trying to relay that excitement and get support for it in a retail market. Most investors today, and going back to 2002, are really rather cynical. They're not interested in the potential for discovery. They'll invest heavily in what looks like a new discovery once it occurs. We haven't, as yet, delivered that discovery hole, that drill hole that announces that a new mine has been discovered. But, we have an enormous amount of geological evidence to let us know we're on that path."

Broughton notes that the Foremore "has a lot to offer, perhaps in more than one deposit style or deposit type. Initially, we were looking for size there, because of where it's located. Road and electricity need to be developed. Smaller high grade findings we know also exist there but we didn't put much emphasis on them. But, now we've got a neighbour there; Nova Gold's Galore Creek is right next door to us, and we're basically halfway between the main highway and where their site is. So, they're developing a road along the top of our Foremore property. And they're going to ultimately develop power across the top of our property; which changes the complexion of everything at our site. They're building it all because they need the access. It's all happening at their cost. And we're going to be the beneficiaries of that."

Roca has also signed option agreements with a group of prospectors to acquire a 100% interest in four separate properties in the Revelstoke Mining Division of B.C. The development of Roca's MAX Molybdenum mine provides a good base for regional exploration of the historically important Lardeau Mining Camp. However, the acquisition of these properties by Roca is partly an act of prudent foresight. "Just to be clear, these are a number of very grassroots exploration projects. What we know is that, generally, once a mine is developed, it's like building a luxury house in a neighbourhood—you increase the price of the surrounding properties. We decided, before we actually go to production, that we would spend some time identifying the best exploration sites and try to acquire them before they increase substantially in value.

"As we're doing the work which will increase the value in the area, we decided to cherry-pick. There were loads of properties available there. So, we have an interesting portfolio, ranging from gold targets, to primarily zinc and lead targets. Clearly the area is pretty rich, like most of British Columbia, and fairly under-explored," says Broughton.

Still, Foremore and Roca's grassroots properties are secondary, understandably, to MAX. "We didn't do much work at Foremore last year because our shareholders wanted us to focus on MAX. We've always said our exploration activities will commence again once we've got cash flow and have a way to finance them that's not so dilutive to shareholders," asserts Broughton. "What we've been working on is an exploration plan at Foremore, which is going to be substantial—many, many drill holes on what we've identified as high level targets that we think have a very high probability of delivering what we're looking for—a mine."

Realizing the benefits of fast-track opportunities, Broughton notes in regards to any future acquisitions, "We are also looking at more and more advanced stage projects and they're not all in B.C., but there's an expectation there to grow and continue the cash flowing side of the business." Though, he says, "I think the results that might be coming out of Foremore, or even MAX, this year will demonstrate that we're also keenly interested in the exploration business."

SULTAN MINERALS INC.

Molybdenum and tungsten are the metals packing a promising punch at Sultan Minerals' Jersey-Emerald property, near Salmo, BC. With multiple mines all rolled into one package of land, including what was formerly North America's second largest tungsten mine, Jersey-Emerald is blessed with molybdenum and more. "It also has lead and zinc," states President and CEO Arthur Troup. "In fact, the Jersey mine, adjacent to the Emerald tungsten mine, was British Columbia's second largest lead-zinc producer."

Under Troup's mindful eye, the Company is following a carefully-crafted blueprint that could lead to success across the board, scooping a whole host of valuable minerals in the process.

With a reputation as a gold, silver and base metals exploration company, Sultan Minerals' presently plans to focus on exploration and development of two past-producing mines. For 2007, a 20,000-foot drill program is advancing to expand resources at the Jersey-Emerald. Cavity monitoring surveys and an environmental assessment of the neighbouring, historical, lead-zinc mine is also underway. In addition, Sultan's other B.C. property, a gold prospect called Kena, is scheduled for trenching and drilling in the near future.

Sultan Minerals, which was incorporated in 1989, is backed by more than 45 years of extensive experience in precious metal exploration as a member of the Lang Mining Group—established by financier Frank Lang, BA, MA, PEng, who is Chairman and Director of Sultan Minerals. As a professional engineer, Lang has been involved in the operation and financing of junior resource companies for over 40 years. He is also President and Director of Cream Minerals Ltd., and Chairman of Acrex Ventures Ltd.

Lang and his former partner, Dick Hughes, shared the Developer of the Year Award for the discovery of the Golden Giant Mine, which was Canada's largest gold discovery at the time, and the first gold mine in production in the Hemlo region in Ontario. "Frank is famous, of course, because of the discovery of the Hemlo gold mine," says Troup, "and he was involved with the discovery of the Federber and Sleeping Giant gold mines that went into production in the Val D'Or area of Quebec.

Mr. Troup has also held positions at many of Lang's other interests—as Vice President, Exploration and Director of Cream Minerals Ltd. since 1987; Vice President, Exploration of Emgold Mining Corporation since September 1993; and Vice President, Exploration of ValGold Resources Ltd. from June 1996 to January 2005. Moreover, as a qualified geologist, Troup is currently Senior Geologist at ValGold.

A graduate from McMaster University in Hamilton, Ontario with an MSc in Geology, 63-year-old Troup has a total of more than 30 years experience in the mining industry in North America, Southeast Asia, and North Africa. He has worked for Rio Tinto, Teck Corporation, Canada Nickel Corporation, Placer Dome Canada Ltd., and the Geological Survey of Canada. From 1990 through 1996, he was President of Archean Engineering Ltd. where, he says, "we would take on projects for these junior companies who didn't have their own exploration staff, and Lang was one of those companies. I managed Frank's exploration in British Columbia, the Yukon, and western Canada for many many years.

"In 1996, Frank asked me if I would come on the board of some of his companies as a Vice President of Exploration. So, I became involved with four of his companies and then, over the years, I stepped down from some of them." In 1997, Lang asked Troup to serve as President and CEO of Sultan Minerals Inc., of which he had been a Director since June 1995.

Troup highlights several of the Company's key directors and personnel, essential to Sultan's past, present, and future success: "Ben Ainsworth is well known, he was with Placer Dome for many years. He was responsible for the discovery of the Howard's Pass lead-zinc deposit in the Yukon Territories. He's famous throughout the industry." Ainsworth graduated with an Honours Degree in Geology from Oxford in 1962 and consequently joined Placer Development as a Senior Geologist in 1965, holding assignments as Exploration Manager, Eastern Canada; Exploration Manager, Chile; and President, Placer Chile S.A. de C.V. His work with Placer included extensive exploration work in B.C. and the Yukon, which did indeed lead to the discovery in 1972 of the world-class Howard's Pass. Then in 1986, he formed an international mineral exploration consultancy and has served on the board of a number of junior mining companies.

"Sarge Berner has certainly worked in the mining industry for many years," says Troup of Sargent H. Berner, BA, LLB, another prominent director at Sultan. Berner has practiced corporate, securities, and natural resources law as a partner in the Vancouver law firm of DuMoulin Black from 1976 until 2004. Berner is a graduate of the University of British Columbia where he received his BA in 1963 and his LLB in 1966. He also received the degree of Master of Laws from the London School of Economics, England in 1967.

"And, Shannon Ross is our Chief Financial Officer. She has a lot of experience as well. She worked with Cominco, many years ago." Ross, CFO and Corporate Secretary, brings more than 25 years of accounting and financial management experience to Sultan Minerals. She began her career in public practice, moved to the mining industry as an internal auditor for Cominco Ltd., and served as a controller and CFO for several mining companies before joining the Lang Mining Group. She holds a Bachelor of Commerce degree and is a registered Chartered Accountant.

Furthermore, Sultan efficiently shares administrational resources at a headquarters in Vancouver, B.C. "There are about 12 people here in the office that are involved in one way or another with Sultan, but not full-time. We have maybe three bookkeepers and two accountants. We have a legal staff that we draw on, but so do other companies."

On the ground at Jersey-Emerald, Sultan also employs seasonal field staff. "There are about seven people out there, right now." Troup says. "Perry and Ed, there's John, Bill, Jack and Spring." As he counts them off, he demonstrates, unconsciously, that he embodies the qualities of a great CEO—one who is constantly in touch with his colleagues; knowing what is happening, and who is involved at all times.

A hive of activity, back in July 2007, Sultan confirmed high-grade molybdenum and the discovery of a new lead and zinc zone at Jersey-Emerald, based on assays received from 13 underground diamond drill holes from a 20,000-foot Spring drilling program. The lead and zinc mineralization is located in the footwall of the historic mine—in one drill hole, this consisted of low-grade mineralization over a core length of 109.67 feet, including a high-grade four foot zone that assayed 11.65% zinc and 6.34% lead.

Three of the 13 widespread holes at Jersey-Emerald were drilled to investigate the vertical thickness and lateral continuity of a molybdenum zone intersected in a prior drill hole that assayed 0.10% MoS2 over a core length of 548 feet. Fortuitously, all three holes successfully intersected the moly zone confirming top grade, excellent thickness, and good lateral continuity. Drill-hole JM07-17 was the highlight of the three holes. It intersected 0.21% MoS2 over a core length of 160 feet. The intersection included a 60-foot interval that assayed 0.41% MoS2, including a very high-grade interval that carried 0.85% MoS2 over 10 feet.

These high-grade intersections that are emerging with each successive drill program have allowed Sultan to continue expanding and confirming the large-size potential of the Jersey-Emerald deposit's tungsten, moly, lead, and zinc, according to Art Troup. No less than 50 holes are planned for 2007. Perry Grunenberg, PGeo, of PBG Geoscience from Kamloops, B. C., is Project Supervisor for Sultan, and a Qualified Person for the purpose of NI 43-101 reporting. The drill program is being run by Ed Lawrence, PEng, who was mine manager for the Jersey and Emerald Mines 35 years ago when the decision was made to halt production.

In a 2005 report on the potential of moly at Jersey, Lawrence wrote that: "This interesting deposit is contained within a large intrusive complex that has been known to be molybdenum bearing for many years. Widespread molybdenum occurrences have been noted on this property since the 1930's. Initially they were found in surface showings on the western slope of Iron Mountain, where the Emerald and Feeney tungsten deposits were developed. In general these occurrences consisted of high grade moly along fracture surfaces, and in the skarns associated with tungsten mineralization."

Thus, Ed Lawrence's report concludes, "A general geological picture of this area is best described as sediments that have been intruded by granitics. A cap of sediments still remains above the area where earlier mining took place."

"Just to give a bit of history on it," says Art Troup. "They also found gold on the property in the 1890s." Records of prospecting and claim staking on Iron Mountain—which owes its name to its large iron-stained outcrops—go back to 1895. The 1896 Minister of Mines Report states that assays as high as 100 grams-per-tonne gold were obtained. They found lead mineralization on the Emerald claims, which was put into production in the early 1900's. And—in the course of mining lead—they discovered tungsten in 1938.

"The tungsten was very close to a lead mine, called the Emerald lead mine. They called it the Emerald tungsten deposit. At the same time, there was a little molybdenite discovered along with the tungsten, but nobody really paid any attention to it. During the Second World War, between 1942 and 1943, tungsten was needed for the war effort, for hardening steel, building tanks and armour-piercing shells and so forth. In very short order, the Canadian government bought the property and put the tungsten mine into production. At that point, it was the largest tungsten mine in the

Western world; production was at 1200 tonnes a day and there was no other tungsten mine in North America at that time."

Post-war the mine remained inactive until 1947, when Canadian Exploration Ltd. (now Placer Dome) purchased the property. Tungsten production recommenced in 1947, followed by lead-zinc production in1949. Troup continues, "Prior to the Second World War, Placer operated an alluvial gold mine in Papua New Guinea, and that was their only mining operation. During the war, the Japanese occupied Papua New Guinea and they were forced out, so the mine was closed. When Placer bought the Emerald tungsten mine, they bid on the U.S. Strategic Stockpile, an American project to stockpile strategic metals. Placer had the only tungsten mine in North America, so they got the contract. With the profits from that tungsten mine, Placer bought Endako and it became the largest molybdenum mine in Canada.

"In the meantime, China started selling tungsten into the world market and depressed the price. But, while Placer was mining the tungsten and exploring for it, they found a very large zinc deposit, which became the Jersey lead-zinc mine, British Columbia's second largest lead-zinc producer," smiles Troup. "Essentially, it's a very mineral-rich area."

He continues, "During development of the lead, zinc, and tungsten, Placer put tunnels underneath these other ore bodies, so they could drive 40-tonne ore trucks in. The ore was mined up above; 200 to 400 feet higher up in the mountain. They would drop the ore down chutes, and then the trucks would pull under and fill up and drive out with it." These tunnels cut through a granite body which was unexpectedly home to encouraging molybdenum values. In fact, there is over 16 kilometres of tunnels that were put in during their operation, and it would cost about $150 million to replicate these workings today. "But, their mandate was to mine tungsten, and the lead-zinc was now at the mining stage. Consequently, while it was interesting to them that this molybdenum was there, they didn't do anything about it."

Placer's tungsten production amounted to 1.6 million tonnes grading 0.76% tungsten, which was produced from the Emerald, Feeney, Invincible, and Dodger deposits over the life of the operation. Lead-zinc production of 8.4 million tonnes grading 1.95% lead and 3.83% zinc was produced from the Jersey and Emerald deposits.

Eventually, though, metal prices became so depressed that mining could barely continue. "The price of tungsten dropped, and lead dropped—lead was no longer used in gasoline at about that time—and zinc went down to

$0.19 a pound. So, the mine was not very profitable. That's why the decision was made to shut the mine down in 1969. They gave the union three years notice that the mine would be shutting down in 1972. They still had some stockpiled ore that was processed in 1973, and then they had a big auction and sold off all the mining equipment."

Only the processing plant foundations were left behind, along with an empty swimming pool—once a popular meeting place for local residents who enjoyed a swim in heated waters from the old mine compressors. The pool still remains a landmark at the site.

The Jersey-Emerald property remained inactive from 1973 until 1993. Following this period of inactivity, Sultan Minerals Inc. entered into an option agreement with the site's owners at the time, Lloyd Addie and Robert Bourdon, to acquire an option to purchase a 100% interest in the Jersey Claim Group. The claims overlie the core area of the former Jersey and Emerald lead, zinc, and tungsten mines that were operated by Placer Dome between 1947 and 1972.

Sultan's interest in the Jersey-Emerald is subject to a 3% Net Smelter Return royalty, which can be reduced to 1.5% by making additional cash and share payments totaling $500,000 and 50,000 shares on completion of a positive feasibility study. Advance royalty payments of $50,000 were to commence in 2000, but amendments to the agreement extended the start of the royalty payments to 2009. In consideration, 400,000 common shares were issued to the royalty holders.

The optioned property initially comprised of 28 crown granted mineral claims, four two-post claims and mineral units, which encompass approximately 1700 hectares in the Nelson Mining Division. The property has since been expanded by staking, optioning and purchasing further claims, to include 47 crown granted mineral claims, 60 two-post claims, and 278 mineral units in 15 four-post claims. To date, Sultan has managed to put together a land package of over 93 square kilometres around the Jersey Emerald property, which includes surface rights to all the access portals, and mine site locations.

Work by the company, conducted since 1994, has defined five distinct deposit types on the Jersey-Emerald property. These are porphyry molybdenum, carbonate-hosted lead–zinc (massive sulphide), sedex-type zinc–silver–copper, and both gold–bismuth skarn and tungsten skarn mineralization.

In May 2005, Sultan also entered into a purchase agreement to acquire the historic Invincible Tungsten Mine. An area of 25 hectares, located approximately six kilometres south of the community of Salmo, within the Jersey-Emerald property boundary. Sultan owns 100% right, title, and interest in the property, subject to a 2% NSR, which may be reduced to a 0.5% NSR by the payment of $150,000 to the seller after the completion of a positive feasibility study, and an Annual Advance Royalty payment of $3,000, which will commence in the year 2010.

As recently as August 2007, Sultan acquired additional surface rights at Jersey-Emerald, entering into a purchase agreement with seller Steve Okipnak to acquire 100% of surface rights to 22 acres of land overlying part of the Jersey claim group. This latest land acquisition takes Sultan's ownership to more than 1,100 acres of surface rights over the proposed mine site, including the access portals to the known deposits and the future mill site as proposed by Wardrop Engineering Inc.

In May, 2007, Wardrop Engineering released its NI 43-101 compliant Technical Report for the Jersey-Emerald property, identifying a potentially commercial tungsten operation at current prices. Following that good news, Sultan is progressing with its 50-drill-hole exploration program as planned, focusing on the site's Dodger and East Emerald tungsten and molybdenum zones. Sultan expects to expand the size and grade of the deposits, thereby improving the economics of the entire Jersey-Emerald project.

So far, four of the holes drilled carry wide intersections of bulk-tonnage-grade moly mineralization, including one remarkable intersection of 0.10% MoS2 over a core length of 548 feet; averaging 0.04% MoS2 over its entire 1,529-foot length. The hole also contained several higher-grade sections including 0.28% MoS2 over 79.9 feet, 0.23% MoS2 over 43 feet and 1.81% MoS2 over 9.9 feet. With rewarding numbers like this to spur them on, Art Troup concludes, "I think we have a really superb property. You know, we have two historic mines; both were very large, at least in their day. And both of the mines shut down, not because they ran out of ore, but because of low metal prices."

Another encouraging fact is that, directly to the south of the Jersey Emerald property, adjacent to Sultan's claims is the 'Molly Mine' still currently owned by Teck Cominco. Although a small mine that was operated in the past, it remarkably yielded over 6% moly.

Certainly then, Jersey Emerald seems to be a 'king' of a property—apt given the history of the company name. "Basically, Frank had a company called Emperor Gold (now known as Emgold) and he was starting another company. He said that he had the Emperor, now he needed the King. But there was already a King mining company, so he said, 'Well, I'll take Sultan.' And that's where the name Sultan Minerals came from," laughs Troup.

And, surely, no Sultan would be complete without a golden treasure trove, which could be Sultan Minerals' Kena gold project.

Widely mineralized and containing several gold, silver and gold-copper prospects, Kena covers approximately 8,000 hectares of land located near the town of Nelson in southeastern B.C. With excellent infrastructure, a power line and major highways passing through the centre of the property, plus a network of new logging access roads, Kena is well-equipped for development. Considerable exploration work had been conducted between 1974 and 1991, before it lay untouched until 1999 when Sultan acquired the project.

Kena's Gold Mountain Zone has a measured and indicated resource of 5,490,000 tonnes at 1.04 g/t, and an inferred resource of 10,710,000 tonnes at 0.967 g/t. The nearby Kena Gold Zone has a measured and indicated resource of 6,330,000 tonnes at 0.969 g/t, and an inferred resource of 1,440,000 tonnes at 1.216 g/t. Despite these very encouraging numbers, along with the incredible opportunities at Jersey-Emerald, the Kena is "not currently our priority," notes President Troup.

"My budget this year for Jersey-Emerald is $2 million, and we have $250,000 earmarked for Kena. That could expand to $500,000 on Kena, if we can get a drill for the property. We're currently doing soil sampling, prospecting and trenching. We have a geologist and a prospector working the claims trying to expand the gold zone to the south, where there exists a low-grade porphyry copper deposit on the property."

Jersey-Emerald has always been recognized for its tungsten and lead-zinc, because it pre-dated the latest wave of interest in molybdenum. However, now that moly is increasingly valued in the world market and in high demand, Arthur Troup notes, "Times have changed." In the early days of work on the property, Troup points out, "They didn't assay moly; they didn't try to mine it. At one point, Placer got into a very rich section and they sent a truck load of the molybdenum ore up to Endako, but that's 1,200 kilometres away. Trucking costs made it prohibitively expensive so it was not investigated further."

But today, the moly is definitely part of the equation for Sultan Minerals. "The mill that we put in will be able to handle lead, zinc, tungsten, and the molybdenum. If we put in a mill that can handle tungsten, we will have all of the facilities required to handle the other commodities. But, if we were going to mine just the lead-zinc, that wouldn't be the case. For the tungsten, we need what's called the gravity and flotation circuits. Therefore, we'll put in a mill that can handle all of the metals."

Troup concludes, "At the present time, we're getting quite a bit of higher-grade moly and we'll be doing more drilling on the moly in the underground. I think we could drill out enough of a resource to justify a 2,000-tonnes-per-day mine." But first, Troup notes, "We have to go through permitting. We expect it'll be fairly straightforward, because the property was historically mined for these commodities, and there are no serious environmental implications from the existing tailings. But, we have to prove up enough ore. Placer always maintained three years of reserves ahead of production—so when the decision was made to close the mine they just mined out the reserves that had been drilled out. We now have to go out and drill out enough reserves to justify putting a mine into production."

Sultan Minerals' CEO says that the company is striving to achieve a minimum of six to eight years of reserves. "We actually have a sufficient tonnage now that we could put a mine into production tomorrow at 1,000 tonnes a day, with a six-and-a-half year mine-life. But, we want to go into production at 2,000 tonnes a day. We need another two and a half million tonnes to justify that operation."

Ultimately, Troup notes, "I think the earliest I could see us getting into production would be in 2010. We're doing a number of different things to prove up a sufficient resource to justify doing a feasibility study. If we can get to that stage within a year, we'll have two years for permitting and mine construction prior to production."

Approaching the deposits "like a layer cake," Troup defines the Company's next steps for mining the ore, which might sound like a perfect recipe for shareholders: "If you think of a wedding cake with three tiers, and the upper tier—the one the bride and groom are standing on—would be the lead-zinc horizon; then the next one down would be the tungsten; and the third tier is the molybdenum horizon. So, you can take each of those tiers and cut them up separately. Same way with our mineralization here; we're on top of a mountain—we can mine it with tunnels. We don't have to sink shafts, we can just put tunnels in from the side of the mountain and drive in with trucks and haul it out. Depending on which level we go in on, we can take out the lead–zinc, or we can take out the tungsten, or we can take out the molybdenum. They're all separately mined."

Troup continues, "I think the tungsten will be first. A tungsten mine would be the easiest to permit and we've got a significant resource now.

There's a low-grade stockpile on the property that we're currently doing metallurgy on, and that might put us over the threshold to make a 2,000-tonne-per-day operation viable. We've already started the environmental studies on that. Wardrop Engineering is doing environmental monitoring of the stream water, and counting the birds and the animals and so forth. So, we're doing all of this and we'll have that information, which we require to start a feasibility study, by the end of 2008.

"The molybdenum is a different situation. The mineralization occurs over a very large area, but much more work is required to determine the size of the deposit. We can only drill local areas from the existing underground workings. We now have to step out and drill test the deposit with some long holes from surface. It's going to take a while until we understand it."

He notes, "It's a really exciting situation. The indications are that the moly could be very, very big—because it has high grade veins. Some of them contain more than 1% molybdenite. It's going to take a lot of drilling until we understand it. It is important to note that Placer Dome put over 5,000 drill holes into the tungsten and the lead–zinc deposits. We have the data from those drill holes, so we know where the deposits go. We really just have to do a little drilling around their holes to prove up the tonnage. So, it's much easier than the moly."

Personally, Arthur Troup says he would like to take the tungsten mine through to production and consider a joint venture with a major for the molybdenum operation, if the size of the deposit merits it. "I'd certainly like to make this one a mine. I don't think there's any question whether it's going to be a mine; it's just how big a mine will it be. I think we have an excellent chance of making three mines out of this property. And I think the molybdenum could be the biggest of the three."

TERRANE METALS

With plans in motion to realize two mining operations in the next five years, Terrane Metals Corp.'s Robert Pease is justifiably pleased with the progress. The President and CEO outlines the steps ahead for these promising, "company-making" properties.

"Terrane Metals story is really born out of the takeover of Placer Dome," explains Pease, a 24-year veteran of the Placer organization. "We created Terrane to complete the transaction to buy these former Placer projects," he says, speaking primarily of the British Columbia-based Mount Milligan, which is a gold-copper project, and the Berg, a property with copper–molybdenum–silver potential. Both seem primed for development, and things have moved quickly since Terrane made the acquisitions in July 2006, with Mount Milligan nearing the final feasibility and permitting stage. "It's all happening," notes Pease. "The company has grown rapidly and we're doing well."

Terrane's portfolio of projects were all previously owned and worked by Placer Dome until Barrick Gold, the largest pure gold mining company in the world, made a hostile takeover bid for Placer in October 2005. By January 2006, Placer had accepted Barrick's price of $10.4 billion (USD). Meanwhile, Placer's Canadian assets, including Mount Milligan, were sold to Goldcorp Canada Ltd. for $1.48 billion (USD).

"I'd been with Placer for many years and, in fact, several of my management team had been involved in the Placer organization also—we knew a lot about Mount Milligan, so we flagged that as being the keystone project," says Pease. So it was that in early 2006, a group of mining entrepreneurs—former Placer Dome professionals whose positions were terminated in the takeover—banded together to make a play for the Placer properties they knew so well.

Pease discerned that the properties weren't a great fit, strategically, for Goldcorp. "Goldcorp is a gold producer and what they really wanted, out of all those Placer Dome assets, was to get their hands on three large operating gold mines in Ontario. So, they got what they wanted, and these other properties, including Mount Milligan and Berg, got thrown into the deal. Basically, we made the argument to Goldcorp that these properties didn't really fit, but had some value and could have a lot more value in the future. What I think I presented to them was a deal at a financial level they understood, allowing them to bank some immediate balance sheet value for these projects, plus some participation in the future up-side value," concludes Pease. "We paid them a preferred share arrangement, so on a fully diluted basis they own 58% of Terrane."

Apart from the accounting benefits of the deal, Pease sold Goldcorp on the idea that Terrane would be able to offer a concentrated focus on these properties, and from a team already well-versed in the background of the projects, particularly Mount Milligan. "In the position I had held at Placer, one of my responsibilities was to oversee Mount Milligan," says Pease. "Since 2002, I was the General Manager for Exploration in Canada, and as well—I had two titles—I was also General Manager of Global Major Projects." Essentially, Pease was responsible for managing all aspects of Placer Dome's Canadian exploration, and overseeing the geological aspects of advanced stage, major exploration and development projects worldwide; just another reason why Pease portrays himself as a convincing leader for Terrane's organization.

"So, in my role at Placer," he says, "Mount Milligan was my responsibility under both titles. That was the other part of the equation I sold them on. Goldcorp is busy on many, many projects, so, I told them, 'I will put a team together, and we will then take these projects and we will continue to advance them and create even more value.' They liked that idea, too. Still, it was very competitive—a lot of people wanted Mount Milligan—and we weren't the only or largest company at the table. But, because of my background with the property, Goldcorp was confident I would be able to put together a really high quality feasibility study on the project, which is obviously what needed to happen."

Terrane picked up four properties in total—Mount Milligan, Berg, Maze Lake (gold) and Howard's Pass (zinc and lead)—for $120 million in preferred shares. Now these projects are part of the story of a rising Canadian mineral exploration and development company, aiming to develop large-scale mineral deposits into operating mines.

Naturally, Mount Milligan is a good fit for Terrane and their ultimate goals for near-term mine operation. Recognized as an advanced-stage porphyry gold–copper deposit, the property is located 155 kilometres northwest of Prince George, in central British Columbia. "A porphyry deposit, a geological term, essentially means a big low grade deposit. They are very amenable to open pit mining development," adds Pease. "Mount Milligan is now the flagship of the company," he continues.

Regionally, Mount Milligan is within the Quesnel Terrane, part of the Intermontane Belt, which was formed of volcanic and sedimentary rocks, between 200 and 300 million years ago. Today, "it's sitting in a site that has excellent infrastructure—roads and towns nearby, power nearby, all those kinds of things—we're not talking about long access roads into pristine wilderness to develop this mine," Rob Pease smiles. "We are talking about an area that is already used extensively, both industrially, mostly for the forestry sector, and by locals for hunting and fishing. Just to point out that it's at the opposite end of the spectrum from pristine wilderness."

Mineralization was first discovered in the Mount Milligan area in 1983, and consequently a total of 194,467 metres of geotechnical and diamond drilling, in 911 holes, has defined a large, gold–copper resource. As of December 31, 2005, Placer Dome reported that the Mount Milligan deposit hosts a measured and indicated resource totaling 205,932,000 tonnes grading 0.6 g/t gold and 0.247% copper. That translates to 3.7 million ounces of gold and 1.12 billion pounds of copper. "Mount Milligan has a history," says Pease, and Terrane is hoping to reap the benefits of acquiring a property previously held and worked by a major for a number of years—including a $50 million database of historical resources.

Prior to Placer's involvement, Mount Milligan was also explored by a variety of companies over the years. Initial regional exploration work was carried out by Selco Inc. and BP Resources Canada Ltd, although, it was a prospector named George Small who was first on the scene at Mount Milligan, pre-World War II. He found gold-bearing float on the western side of the property, but his investigations were interrupted by the war. Small tried again in 1945 but was unsuccessful in making any further progress.

Twenty-seven years passed before Pechiney Development Ltd. entered the fray; completing induced polarization work, geochemistry, and drilling five holes at Mount Milligan. But their claims lapsed and they left empty-handed, too. The property was overlooked until 1983 when Selco arrived and began their program of extensive geochemical work which identified significant copper–gold mineralization in what is now known as the Main deposit. In 1984, Selco amalgamated with BP and that same year another prospector, Richard Haslinger, staked claims on adjacent ground to Mount Milligan, which BP then optioned. BP launched a program that included extensive geological, geochemical, geophysical and trenching work.

BP's program continued until 1986, when Lincoln Resources Inc. optioned the property. Lincoln discovered significant copper–gold mineralization in the Main deposit. The company later reorganized to become United Lincoln

Resources Inc. and then amalgamated with Continental Gold Corp., which uncovered Mount Milligan's Southern Star deposit in 1989. This work also identified low-grade porphyry gold and copper–gold mineralization in the Goldmark and North Slope zones.

Ultimately, Placer Dome purchased BP's interest in the mineral claims and acquired Continental Gold in 1990—against stiff competition from mining powerhouse Noranda—for $230 million (USD). Placer thereby gained control of the Mount Milligan claims. They completed a pre-feasibility study in 1991, and in November 1993 a Mine Development Certificate was issued for the Mount Milligan project, effectively authorizing Placer Dome to construct and operate a mine on the property. But low copper and gold prices delayed mine development, and Placer's certificate expired ten years later, in November 2003.

"Copper prices had really slid off—we essentially went through a decade of copper prices that were very, very depressed. It made the economics of a project like Mount Milligan very questionable," Pease speaks of the rock-bottom metal prices in the early 1990s. "Part of the thinking, too, at Placer, was recognizing that in an environment with good copper and gold prices, Mount Milligan is a great project. It just needed to wait for the right time to come into production. So, Placer thought, 'We'll sit on that till the time is right.' And they were right to wait as we've now seen, metal prices improved in 2003."

In 2004, Placer Dome was encouraged enough by the turning metal prices to conduct a 14-drill hole twinning program to obtain bulk metallurgical test-work samples representative of the MBX Zone, 66 Zone, and Southern Star ore-types. "Yes, we were working on the project again," explains Pease. "We commissioned a large engineering firm, AMEC, to update a feasibility study on the project." Also, metallurgical test work was re-started, a 3D geological model was redefined, the pre-feasibility study was initiated, discussions with First Nations and local communities were reactivated, and discussions began with the government regarding re-permitting the project.

Evidently, Placer Dome viewed Mount Milligan as a worthy project, although the AMEC study was never completed due to the Barrick take-over. "Certainly, we try to impress on people that: no, we didn't just inherit a project that was locked away in a vault with three inches of dust on it," says Robert Pease of Terrane Metals. "This was really being worked on again, and was obviously a project of interest to Placer. So, we were then able to do the deal that created Terrane, get the project over into Terrane, and then put my team together so that now we've launched a full feasibility study, and we've been working away on that, and we're on track to have that feasibility study done by early 2008."

The upcoming feasibility study, to be conducted by Wardrop Engineering Inc., follows Terrane's multi-rig diamond drilling program that collected bulk samples for metallurgical test work. The Phase I drilling program consisted of thirty-two holes, comprising 8200 metres of large diameter core drilling, designed to provide representative bench and pilot plant test-work samples over a range of gold and copper grades. Notably, several of the holes in Phase I intersected significant thicknesses of laterally continuous gold–copper mineralization outside of the historic resource limits. Terrane's Phase II drilling involved another 8,200 metres of core drilling in 22 holes. This program was designed to upgrade Inferred resources into Measured and Indicated categories.

Reflectively, Pease ponders for a moment and then says, "I think we're roughly on schedule. We have released a new resource estimate for Mount Milligan, which we are very pleased with. The deposit now is estimated to contain measured and indicated resources of 417.1 million tonnes grading 0.21% Cu and 0.41 g/t Au or 1.9 billion pounds of contained copper and 5.5 million ounces of contained gold. We've done an awful lot of metallurgical test work as well. That's gone very well. So, now it's a case of how much of that resource are you going to be able to recover, essentially. And then the full feasibility study will be done in early 2008." Collectively, all these elements put Terrane in positive stead for their future plans of a fast-tracked mine operation.

The company expects to have all mining permits and certificates in place by late 2008. Once the feasibility is complete, financing will be arranged, for the construction costs of $800 million (CDN). "Say we get our permits by late 2008, hopefully then we'll be in a position to start construction right away. Construction will probably take two and a half years minimum, to get the mine built and operational. It's a large, industrial construction project," says Pease. Ultimately, Terrane foresees the mine to be producing by early 2011, with a projected throughput rate of 60,000 tonnes per day and a predicted mine-life of 15 years.

The construction phase itself is expected to be a boon for local employment, as up to 900 people would be hired to build, while 300 to 400 people will also be working in the mine for over a decade afterwards. Additionally, in terms of First Nations, when he was Placer's Project Manager back in the 1990s, Pease established a conscientious relationship with First Nations groups in the Mount Milligan area. He is furthermore ensuring that interested parties can participate in the ongoing exploration and mining activities.

Thus, Terrane's upcoming plans arguably place them in a strong position for realizing their mining objectives. "Yeah, because we really are close," says Pease, animatedly. "Though it is kind of 'sexy' in the market sometimes: the discovery. I'm a geologist—I get caught up in all that. Discovery is fun. But the reality is that you start to look and say, 'When are these discoveries ever going to be mines? When are they going to generate revenue?' There's a whole bunch of risk in it, a whole bunch of uncertainty, and that's a timeline you probably can't even realistically deal with.

"Permitting is our next hurdle. There are a number of milestones but we hope by late 2008 we'll have our permits. When we talk about mine permitting, essentially, we're really talking about mine certification, then you can get the detailed permits you need to operate the mine. Because Placer's certificate eventually ran out in 2003, we knew the project had to be re-permitted. But, there's no clause in this new permitting process that if you've been permitted in the past then you get a fast track on a new permit, bypass such and such and head straight to 'Go'. We still have to move through the process as any project applicant would, but our benefit is that we have all of that baseline data and all of the background studies that were done to get a permit originally," Pease adds.

"All that data is still valid, so we have a really good place to start from. And, of course, we understand the project very well since it has been studied extensively in the past, so we're not really breaking any new ground," notes Pease. "Yes, the permitting process has changed a bit since 1993, but fundamentally the basic requirements—what you need to do in terms of baseline studies—are still fairly similar. There's some additional things that need to be done, and we're doing those right now, and so, like I say, we're on track.

"One of our big milestones," Pease explains, "will be submitting our environmental assessment document. We will do that just prior to finishing the feasibility study, so we're hoping we'll have that submitted in early 2008." Once these scheduled hurdles have been overcome, Pease is aware that there will be a "major shift." He says, "Right now, the load is on us—to perform all the studies, write all the reports, do all the work. Then, we've proverbially thrown the ball over the fence and it's all up to provincial and federal agencies to review your work and decide: Does it meet their guidelines and regulations? Is it acceptable? Looking at socioeconomic concerns, what have you done with First Nations— there's a long list of criteria. And, if you have met the criteria you can anticipate getting approval."

While Terrane crosses the t's and dots the i's in preparation for Mount Milligan's next big steps, fortunately they haven't forgotten the possibilities of one of their other properties, the Berg, also based in B.C. "Berg is a little less advanced, but it's our molybdenum asset. It's a porphyry deposit, south of Smithers, containing copper and moly—a fairly classical deposit. There are good descriptions in the geological literature on it. It's a project that saw a lot of exploration work in the 1970s and into the 1980s."

Kennco Explorations (Western) Ltd. first recognized the potential for porphyry copper style mineralization at Berg in the early 1960s. Their work demonstrated that the Berg exhibited deep effects of surface leaching and revealed the widespread presence of 'supergene' minerals—usually a result of weathering that oxidizes the primary (hypogene) sulfide ore minerals and redistributes the metallic ore elements. This is particularly common in copper ore deposits. During 1965 and 1966, Kennco's drilling delineated two main mineralized zones. The northeast zone contains hypogene and some supergene mineralization, while the southeast zone contains widespread supergene mineralization.

In 1972, exploration and development of the property was taken over by Canex Placer Limited (Placer Dome) under agreement with Kennco. By 1980, a total of 146 diamond drill holes totaling 18,593 metres had been completed on the property. Placer Dome also managed to complete numerous resource estimates, metallurgical tests, environmental studies, and financial analysis on Berg. Placer Dome had defined a geological resource, with a 0.25% copper cut-off, of 238 million tonnes grading 0.40% Cu, 0.052% MoS2 and 2.84 g/t silver.

"Yet," Pease notes, "not much has been done since the 1980s, because of low copper and molybdenum prices." In fact, the only activity on the project was the digitization of the drilling database by Placer Dome in 1992. "Also," explains Pease, "at that time Berg was a joint venture between Placer Dome and Kennco's successor, Kennecott, and the deal didn't really work for either of the partners. Berg was controlled and operated by Placer, who owned 51%, while Kennecott owned 49%. Placer Dome, by 1980, had made their decision to focus on precious metals – they weren't that interested in a copper–molybdenum project without a significant gold credit. So, while both parties recognized Berg as a very interesting resource, a serious deposit that would be worth something one day, it was just not strategically important enough at the time for them to spend the money necessary to advance it."

Now that moly prices are riding high, and forecast to stay high or rise even higher, and copper prices are up as well, Terrane has secured 100% of the Berg's eight claim units and one mining lease. The property, located in the Tahtsa Ranges of central B.C., is approximately 84 kilometres southwest of Houston and 585 kilometres north of Vancouver. "We acquired Placer's 51% from Goldcorp, and then we very quickly had discussions with Kennecott. In September 2006, we bought out Kennecott's 49% interest," says Pease. "We paid Kennecott $1 million (USD) and granted them a 1% NSR royalty, which means they will earn 1% of the revenue from the mine once we go into production."

Furthermore, it is noteworthy that the Berg property lies in close proximity to the Huckleberry Mine, an open pit copper–moly mine. "Huckleberry is 22 kilometres away—of course, that mine wasn't there back in the 1970s and 1980s when Placer Dome and Kennecott were spending money on the Berg project," says Robert Pease. "Berg was a little more remote in those days. But now we have an operating mine only 20 kilometres away. We believe that project has a lot of synergy value. It's known that the Huckleberry Mine has very limited reserves left—it's running out of ore, if you will. Having that infrastructure close by could be significant for our development."

Berg's development is farther down the road than Mount Milligan's, but "not by much," notes Pease. He explains that the manpower is being made available to advance the Berg project. "The Berg camp is built for 20 or so people, and we started a major drill program in the Summer of 2007: two drills with a crew of four for each drill, plus a foreman; two to four geologists at any given time; four to six technicians and samplers; plus a cook and camp manager. Berg has that historic resource, that's documented. What it needs is a new modern drill program to update and expand that resource."

Terrane has commenced that work with a 12,000 metre diamond drilling program at Berg, focusing on the historic resource area, which is believed to be open to expansion. The drills are testing for continuation of the mineralization both laterally and at depth. Diamond drilling is expected to be completed by the end of 2007. "We're expecting drill holes that will have continuous long intersections with good copper and molybdenum mineralization." Pease says, "We will re-prove the original drilling, and acquire the data to accurately model the deposit."

Terrane has commissioned Wardrop Engineering to undertake a Preliminary Assessment of the Berg project, and has engaged two additional consulting engineering firms to undertake studies in support. AMEC Earth & Environmental is under contract to initiate an environmental baseline campaign including water quality, hydrology, meteorology, waste–rock characterization, wildlife and vegetation studies. BGC Engineering will provide geotechnical services to evaluate waste–rock management and open pit design. Wardrop Engineering will use these studies, in conjunction with their new resource estimate, to complete an NI 43-101 compliant Preliminary Assessment of the project in early 2008.

In addition to the extensive work being done to advance the Mount Milligan and Berg properties, Terrane is also working, albeit at a lesser scale, on the Maze Lake and Howard's Pass properties, which were also part of the parcel acquired from Goldcorp.

Maze Lake is located in the Kivalliq District of Nunavut, 90 kilometres southwest of Rankin Inlet and 45 kilometres west of Whale Cove, both coastal communities on Hudson Bay. The project consists of five Inuit Owned Lands Mineral Exploration Agreements (EA's) with a total area of 39,866 hectares.

Only minimal exploration work has been done in the area, but in 2000, WMC Resources Ltd. outlined a 30 kilometre long gold anomaly. Follow-up work to this initial discovery defined further anomalies. Maze Lake was acquired by Placer Dome in 2003 and exploration-drilling under the gold mineralized boulders on the property intersected 1.35 g/t gold over 16.5 metres in the first drill hole and 6.29 g/t gold over 4.8 metres in the second hole. The second intersection included a higher grade interval of 15.06 g/t gold over 1.9 metres.

Terrane's current plan for Maze Lake is to option it out to Laurentian Goldfields Ltd., a Vancouver-based, private gold exploration company. "Laurentian will have a right to earn a 75% interest," explains Pease. "The main reason to do that is strategic. It's a good project—a potentially high-grade gold deposit—but it's still at an early exploration stage. We've done some drilling and, yes, there's some gold there. It's encouraging. But to advance the project and get better value for Terrane shareholders in the long term, it makes sense to place Maze Lake with a company that can focus on it as one of its primary properties."

The Howard's Pass is the last of Terrane's properties, located along the Yukon–Northwest Territories border. Between 1972 and 1982, Howard's Pass was a joint venture between Placer Dome (51%) and US Steel (49%)

and was identified as a zinc and lead resource. In 2005, Pacifica Resources Ltd. secured an option to purchase a 100% interest in Howard's Pass. Pacifica, which was re-organized and re-named Selwyn Resources Ltd. in 2007, was required to make $10 million in option payments over a seven year period to the original joint venture partners. In addition, Selwyn was required to undertake $3.5 million in work commitments, which it has completed, and to grant the partners a 20% net profits royalty (capped at $10 million) and a 1% NSR royalty.

"Howard's Pass is a very interesting project," notes Pease, "it's one of the largest known zinc resources in the world. It's really a very large zinc deposit. Nobody knows how big it is. It also has high amounts of lead in it and literally runs for kilometres and kilometres. The zinc people of the world have always been interested in it; pounding on our door. I actually negotiated the agreement with Pacifica, when I was at Placer." After Placer Dome was taken over by Barrick, Terrane Metals Corp acquired Placer's 51% interest in this joint venture from Gold Corp. Terrane will therefore receive Placer Dome's share of the payments and royalties from Selwyn. "So presently," Pease explains, "Terrane owns 51% of Howard's Pass, under agreement that Selwyn will ultimately own 100% of it."

Terrane's strategy is to manage its priorities to maintain focus on its most important properties: Mount Milligan and Berg. And this focus is motivated in large part by current copper and moly prices—"Very much so." says Pease. Moreover, in contrast to other mining companies that are content with developing a resource and then selling the property on, with no intention of owning an operating mine, Pease has made himself crystal-clear that Terrane definitely intends to follow through until these projects are in operation. "Yes, yes—these two assets that we have, Mount Milligan and Berg—they're both very significant deposits and mineable. Our intention is to do absolutely everything we can to develop them into mines.

"If everything falls into place, with the current scheduling, Mount Milligan would be coming into production about the same time that we are building Berg," says Pease.

"So, here's Terrane Metals," Pease notes enthusiastically, "with two very good-sized, significant deposits, both on track to production. It's our intention to build a company out of them. These are 'company-making' deposits—big deposits that are going to generate large amounts of cash flow. They present a real opportunity to build a company—not in two decades, but in five years. We have a five year plan here, where we can have two mines in production."

Rob Pease, PGeo, FGAC, is joined by an impressively qualified management team—including several Placer Dome geologists—who are all familiar with and excited by the prospects of Terrane's two key properties. As Pease highlights, "There's hundreds of junior mining companies to invest in, so I suggest looking at the management team—we're all experienced mining professionals in our areas of expertise, and we're all dead serious about this."

Terrane's team includes Darren O'Brien, PGeo, as Vice President–Exploration. O'Brien was employed in the Placer Dome group for the past 12 years, most recently as Senior Geologist, and prior to that as Project Geologist on the advanced stage Pueblo Veijo project in the Dominican Republic. Terrane's Chief Financial Officer, Edward Farrauto, CGA, has over 19 years experience as a senior financial officer in private and public companies. Peter Marshall, PEng, is the Vice President–Project Development. Prior to joining the Terrane team, Marshall was employed by Placer Dome for 19 years, in a variety of positions including Construction Project Manager for the Cortez Hills project in Nevada. Terrane's Vice President–Corporate Affairs and Sustainability is Glen Wonders, RPF, MBA. Wonders, a Registered Professional Forester (Alberta and British Columbia), earned his BSc in Forestry from the University of Alberta in1985, and his MBA Royal Roads University in 2005.

Likewise, the Directors on Terrane's board are equally experienced and noteworthy: in fact, directors Douglas Leishman and Jeff Franzen worked closely with President Robert Pease to form the company. Leishman, PGeo, FGAC, ARSM, is a geological consultant. He was formally the Director of Geology and Exploration for Endeavour Financial Ltd., a financial advisory firm focused on the mineral industry, and a Senior Mining Analyst with Yorkton Securities Inc. in Vancouver. Jeffrey Franzen, MSc, PEng, has over 30 years experience in mineral exploration, mine development and operations, and is a registered professional engineer in B.C.

"And then, there's John Reynolds" says Pease. The Hon. John Reynolds, P.C., who has served as both an MLA in B.C. and as an MP in Ottawa. Prior to his recent retirement from federal politics, Reynolds was the Official Opposition House Leader for the Conservative caucus. Previously, he had been Leader of the Opposition in the House of Commons for the Canadian Alliance and is currently a Member of the Queen's Privy Council for Canada and a Senior Strategic Advisor for the law firm Lang Michener LLC.

"It has surprised me a bit, but John's been very engaged." Pease says, "He's very interested in the mining business and fundamentally believes that mining is a big part of the backbone of the economy of this country. Mining is such an important thing—helping in rural parts of Canada and helping with some of the problems First Nations have. John thinks mining is so much a part of the solution to a lot of those problems. So, as a director he's very engaged. John asks lots of really good questions, provides excellent advice, and it doesn't hurt to have a guy on your board who can pick up his cell phone and call the Prime Minister," laughs Pease.

In good spirits, Pease is keenly aware that it can take more than just connections (to Placer Dome, the Prime Minister, or otherwise) to make a company successful. Fortunately, with the advancement of two potential mines, Mount Milligan and Berg, Terrane seems to have assembled the right ingredients.

Rob Pease concludes, "We're still a developing story. But a lot of the things that are involved in the true 'value creation' in this company are on schedule to happen, are happening, and will happen—so I believe there's a real future in front of Terrane."

THOMPSON CREEK METALS COMPANY

The new Toronto-based offices of Thompson Creek – the recently re-christened name of 'Blue Pearl,' one of the world's largest publicly-traded pure molybdenum producers – are befitting a company that has ascended high above many of its junior competitors, to become one of the best in the business.

Twenty stories high, Executive Chairman Ian McDonald has a bird's eye view of the hustle-and-bustle of Bay Street below. He can even see his old headquarters, nestled among several other skyscrapers on Adelaide Street, and he points down at the building from his floor-to-ceiling window.

Their recent relocation is just one of many changes, McDonald notes, as he throws away his old Blue Pearl business cards in preparation for the freshly-printed Thompson Creek cards that are en route. "Thompson Creek is well known in the molybdenum business, so that's what we're changing the name to," McDonald explains. "It has more than 700 employees and is an established entity; known throughout the steel industry, and in the United States. We're going to become more American, while the head office remains here in Canada."

The strategy to Americanize the company is expected to attract U.S. investors once they are listed on the New York Stock Exchange as well as the Toronto Stock Exchange.

Not that publicly-traded Blue Pearl wasn't already a name in the molybdenum business in the few years prior to completing the US$575 million acquisition of privately-owned Thompson Creek Metals Company, an integrated moly producer, in October 2006.

Nevertheless, the transaction transformed the former junior mining company into the world's fifth largest moly producer. Ian McDonald was Chairman and CEO at the time of the acquisition. "It was a transmogrifying deal," he says. "To take Blue Pearl - essentially a small development company with only three full-time employees, including myself - and we turned it into Canada's ninth largest mining company."

Along with the Thompson Creek open pit mine and concentrator in Idaho, the acquired assets included a 75% interest in a second open pit mine, concentrator and roaster in Endako, British Columbia, and the Langeloth metallurgical refinery plant in Pennsylvania.

Meanwhile, Blue Pearl was also developing its own molybdenum project, Davidson, a high grade underground molybdenum mine, near Smithers, B.C., about 200 kilometres northwest of the Endako complex.

Needless to say, it was a time of vast growth for Ian McDonald and the Blue Pearl team. "We were in the office over there for a few years, on Adelaide. Sharing with Glencairn (a gold mining company), but it was time for us to go out on our own. We were sharing people to save costs and now we needed our own space," says McDonald, casting his eye around his sky-high corner office, just down the hall from a mahogany-paneled reception area, complete with an executive pool table. "Nothing too extravagant," he says. The company also has offices in Denver, Colorado and Vancouver, B.C.

Prior to his appointment as Executive Chairman, McDonald had held various positions, including Chairman, President and CEO since the company acquired the Davidson property in 2005. But, his direct involvement in the mining industry began back in 1987 – when he literally struck gold with Glencairn and Wheaton River.

"I spent 20 years in the gold business. I was one of the founders at Glencairn Gold Corporation; Chairman of the Board from 1988 to 2006. I was also at Wheaton River for 15 years; I was founder, first shareholder and Chairman and CEO for 11 years until 2001," McDonald says, understandably proud. Wheaton River Minerals Ltd. is the first and only successful Canadian heap leach miner and is now part of Goldcorp Inc., listed on the New York Stock Exchange. "I stayed on the board while we did another transforming transaction with that company. I can't take credit for the real increase in size, but we did have one of the only successful gold mining companies – certainly of the smaller companies – in the 1990s."

McDonald continues, "I've always been a developer and a miner, both at Glencairn and Wheaton River. We had mines, we produced and sold gold, and we had employees."

Overall, with more than 25 years experience in the resource sector, the 52-year-old native of Deep River, Ontario, is a graduate from Toronto's Ryerson University Business Program. "I graduated in 1977; been in the mining business ever since - mining and finance business from 1981 to 1988, and full-time running public companies since 1988."

After his graduation, he joined Richardson Greenshields (now integrated into RBC Dominion Securities) in 1981 as an investment adviser and a specialist in derivatives. In 1986, McDonald moved to Yorkton Securities

Inc. as an investment adviser, specializing in mining issues, and built a solid repute as one of the firm's leading investment advisers for medium to high net worth clients.

This led to his successful gold mining career, and to his present position as Executive Chairman with a leading molybdenum company. "I've only been in the moly business for three years, and what a three year period. All I can say is, we were lucky. We had no idea it was going to be this kind of ride."

McDonald says, with enthusiasm, "I love business – I'm not that creative, in terms of: I can't write a song or a poem. However, I am good at structuring a deal. I love the 'deal' part of business. I just get such a kick out of it."

Certainly, it's not a prerequisite of the job to enjoy what you do, yet it clearly helps that McDonald is passionate about deal-making. And, furthermore, he's enjoying Blue Pearl's/Thompson Creek's rapid rise to its current status as a moly company with a $1.9 billion market cap.

"In 2004, the company had a total value of $1 million – today, the company's worth $2 billion. We created that wealth by doing smart things, and that is such a thrill," says McDonald.

Now adopting a universally-renowned name in molybdenum, there is little doubt that Thompson Creek is doing many 'smart things' to secure its fate as a world leader. In fact, the company annually produces 5% of the world's molybdenum supply - in excess of 20 million pounds of molybdenum.

Moreover, it's noteworthy that Thompson Creek is the second largest open pit primary moly mine on the planet; located 35 miles southwest of the town of Challis, in Idaho's Custer County, a historic mining area. "And the Endako mine, in B.C., is the world's fifth largest primary moly mine," says McDonald.

Endako is a joint venture between Thompson Creek and Sojitz, a major Japanese trading company. "Sojitz is the largest molybdenum trader in Japan, and they're the second largest steel trader in Japan as well. A very big trading house," explains McDonald. "They have a 25% stake in Endako (Thompson Creek holds the other 75%), and we're discussing them buying 25% of Davidson."

The Davidson Deposit, just outside of Smithers, in B.C., was Blue Pearl's first property in the moly business – a high grade, underground deposit lying under Hudson Bay Mountain. Molybdenum was discovered there way back in 1944. "The global resource is bigger at Davidson," says

McDonald, "but it's an underground mine. There won't be as many tons per day. Not near as many. We're just finishing up the feasibility which is based on 2000 tons of very high grade that we will process at the Endako mill – we'll just truck it 200 kilometres down the road; it's right on the highway, right on the rail road too. It's going to be a good mine."

Once developed, the Davidson Deposit will be one of the highest grade molybdenum mines in the world.

Meanwhile, the Langeloth metallurgical plant in Pennsylvania is the larger of Thompson Creek's two smelting or roasting facilities. With a roasting capacity of 35 million pounds, the plant houses six multiple-hearth roasters for the conversion of concentrates to technical molybdenum oxide. It also has capacity for ferromolybdenum, pure molybdic oxide and other moly materials. The complex also has a plant for toll recycling of spent catalysts.

"What we do is we take the concentrate from the Idaho mine - most years, it's 15 to 20 million pounds. Langeloth can do 35 million pounds, so whatever we don't use in the excess capacity, we fill it up with either toll recycling or custom roasting for other mines... We also have a smaller plant in Endako. So, we produce the finished product in Endako as well. All it means is we're not at the mercy of somebody else's roaster or smelter."

Thompson Creek's metallurgical facilities at Langeloth and Endako currently have the capacity to produce 50 million pounds of technical grade molybdenum oxide per year, which is approximately 12% of the world's total roasting capacity.

If that fails to impress, then consider that Thompson Creek's Idaho and Endako mills together process up to 56,000 tonnes of molybdenite ore per day. (Molybdenite contains approximately 60% molybdenum) Whichever way you look at it, Thompson Creek is a substantial mover in the moly game.

"We're a big player there. These are awfully large holes in the ground – Endako and Thompson Creek. They're both predictable, long life mines. Before we took over in October, they were owned privately and they went through some pretty tough years - they've been raking it in the last couple of years, but they did have a few lean years. Actually, they were mining without geologists even on staff, which is truly unheard of.

Thompson Creek has total measured and indicated mineral resources in excess of one billion pounds, including 246 million pounds of proven and probable mineral reserves (according to Blue Pearl's 2006 Annual Report).

The mineral resources and reserves are currently being re-evaluated using a higher molybdenum price and updated costs for the development of new mine plans at Thompson Creek and Endako.

Wardrop Engineering Inc. has been hired by Thompson Creek to examine the feasibility of constructing a super-pit at Endako, unifying the three existing pits into a large single pit, and of increasing mine production to 50,000 tonnes per day from approximately 30,000 tonnes per day, with a proportionate increase in roasting capacity.

These new plans will be based on a re-calculation of mineral resources and reserves assuming a long-term molybdenum price of US$10 per pound and updated costs. Previous mine plans had assumed a long-term price of $5 per pound at Thompson Creek and $3.50 per pound at Endako.

"We're using $10 per pound as a base cost, long-term going forward. I'm actually way more optimistic than that, but in the mining business you better hope for the best, but plan for the worst," notes McDonald. "You look at all the feasibilities of all the other juniors out there, even the most optimistic ones are $4 to $7 per pound. First you have operating costs, and then you have to get your capital costs back; so, you're at $10 without making much money. That's why I don't think you're going to see the new go-forward price below that," McDonald concludes. "But, you never know, right?"

Once the re-evaluations are complete, the new mine plans are looking towards a long and prosperous future for Thompson Creek's properties. "The mine plan at Endako is going to be around a life-span of 20 years. We haven't even done any more drilling. If we do more drilling, I'm sure it will go further. It's been in production for 42 years; I'm quite sure that when that mine shuts down, I won't be living and breathing on this planet. It's just the nature of the orebody," laughs McDonald.

Similarly, the Thompson Creek mine in Idaho has been in production since 1983. "It's going to be there decades as well," says McDonald. "Essentially, you always want to keep 10 or 15 years ahead of you, also our shareholders want us to. It's a lot different from the gold business, when you want to have that magic number of seven years ahead of you."

Thus, with a healthy projected life-span and strong growth potential at Thompson Creek's two operating open pit mines, production is expected to increase from 21 million pounds in 2007 to 27 million pounds in 2008, and subsequently 29 million pounds in 2009 – a sizable piece of the global moly pie.

"Today, there is an approximately 420 million pound world market, and we're hearing about this 'molymania' - which is a nice catchphrase," says McDonald. "But, the reason that we're in a real sweet spot in the cycle is because there was an underinvestment in exploration and engineering for molybdenum deposits for the last 25 years. The price was low - US$3 per pound compared to $28 and $30 per pound now. And China was swamping the market – exporting 100 million pounds in a 230 million pound market, in the 1980s.

"But, now, China – I don't want to say 'they've got religion' but they are keeping and using more of their own molybdenum. And since they're still exporting it, I think they're more interested in higher prices. They've become more of a capitalist economy, in a lot of ways, over the last few years. They want to create some world class mining companies. So, what they've done is taken more control over the 'mom and pop' operations, and I think they want people to manage resources in a more efficient manner. That's the real reason we're now getting discipline on the part of the Chinese to manage their resources and enhance the value of commodities. It happened with Zinc, now it's happening with molybdenum."

Also contributing to a likely spike in molymania, McDonald says there is a dawning realization of molybdenum's seemingly endless potential. "Molybdenum is not like some other commodities like copper or gold, where you just sell it. With molybdenum, you actually have customers, you have to manage that process, and then the product we sell varies; it's not a homogenous product. It has various applications."

Used primarily as an alloying agent to enhance the strength of steels employed in harsh or high temperature environments, molybdenum can be found in airplane engines, nuclear power plants, oil and gas pipelines, offshore drilling rigs, building exteriors, and tools. Furthermore, moly is an important alloy in high quality stainless steel and as a catalyst for removing sulfur in petroleum as part of the oil refining process.

The company additionally produces 'value-added' products such as ferromolybdenum (FeMo), a key ingredient in steel and cast iron manufacturing. The Langeloth facility is the largest producer of FeMo in North America.

Thompson Creek's Ian McDonald reaches into a closet behind his desk and pulls out several varieties of molybdenum for a brief 'Show and Tell.' "High grade molybdenum is a blue-grey noble metal; that's where the 'Blue' in the Blue Pearl name is derived from. We sell the oxide either as powder or brickettes. Then, this is ferro – it's heavy because it's got iron in it; we're the biggest makers of that in North America, second biggest in the world."

McDonald puts away the rocks, yet the moly isn't quickly forgotten; leaving behind its signature graphite-like silvery residue on the hands of those who touch it. Not that it's likely that Ian McDonald will soon need reminding of the metal that is making millions for his company.

"We're very, very fortunate," says McDonald. "In hunting, they always say 'one in the hand is worth two in the bush'; this works as an analogy within the mining business. One built mine is worth four or five that are on the drawing board. Basically, it takes a long time to build a mine."

McDonald continues, "No matter how good the projects are, there's always some completion risk. These are not like factories. Every single orebody is a little bit different than all the others. Whereas in a factory you can put more money in it, put more raw material in one end, and eventually; more widgets come out the back end. But you can't usually do that with mining – it's a living, breathing thing.

"Every day you start mining, there are problems and issues that have to be addressed. There's a certain size that a mine is pre-destined or optimized at. You can't just make it bigger because you have more money and you want it to be bigger."

The geometry of the ore body can also be crucial. "A gold mine, for example, might have a three meter wide vein, so you're not going to be able to extract two to three thousand tonnes a day. Yet, with the big open pits you have more flexibility."

McDonald speaks from experience – something that puts Thompson Creek at a distinct advantage over most rivals. "Look, we're operators, and we've built a couple of mines in the last 10 years," McDonald points out. "We know what one looks like, that's the main thing.

McDonald admits, "In this day and age, there are really high barriers for any new company to build a mine. Because there are no people. You couldn't just go find a great ore body and hire people and start mining. It can't happen. Everybody in the mining business has a job. So it would take a while to get the personnel. On the environmental front, it takes a long time to get things permitted. It's really hurt the industry. But you know what, that's just the way it is. That's the new reality.

"I don't want to rain on anybody's parade but that's the challenge the other guys face. Getting them built is the challenge."

In the near future, permitting will be on the horizon for Thompson Creek as the company looks to expand both of its open pit mines, while Davidson will require permitting for the entire operation. Ian McDonald discusses how he foresees the often arduous permit process proceeding for these cases, "Davidson is much more simplistic than most mines, because we're just hauling it away (to Endako). And molybdenum is a very benign element – it's in your vitamins. You can't live without it. Plus, the Davidson ore body is actually acid consuming – it's got very good mineralogy. So, it should be simple to permit.

"For Endako, we will need to permit if we expand and we're looking at that. We need to get a permit when we increase the reserves at the Thompson Creek mine because we need to put a lift on the tailings dam that is on site. (Tailings dams are used to contain tonnes of mineral processing industry wastes.) The dam is twice as high as where we're sitting right now, on the 20th floor." McDonald glances out of his window – it's a long way down to the road below. "The dam is 500 feet, and we would propose to take it to 600 feet. It will be the second largest earth and tailings dam in the world. We will need to permit it, but there's no reason we can't. That's why we have engineers, that's what mining companies do."

Another angle that will be on the mind of McDonald and the board at Thompson Creek is the environment. With mining companies occasionally being perceived as 'bad' for the Earth and only caring about the cost to their bottom-line, it might seem refreshing that McDonald and his associates are conscious of the animals and ecology in and around their properties, too. "We're not destroying the environment, we're good stewards of it," says McDonald. "At the mine sites, these are a few thousand acres we control; these are animal refuges. During hunting season, the deer just come piling onto the mine site – as they somehow know they're not going to be shot at, so they're safer there!"

Moreover, while at Wheaton River, McDonald permitted the first heap leach mine in Canada, and in Costa Rica – infamous for being tough on corporations working in its highly cherished and protected eco-diverse regions – McDonald managed to get a permit for a Glencairn gold mine. "Everyone said we'd never get it permitted – and now we run 'eco tours' on the site, and we employ 10 game wardens to prevent poaching at that mine."

Of course, working in harmony with local communities and Aboriginal tribes is also important to achieving success in the mining business. "We will be signing a participation agreement with the local band at Endako, B.C., they're called the Witsuwit'en. We do not presently have an agreement with them, because the mine was built before those existed. I've dealt with several bands in B.C. and in the Yukon. There is no land claim settlement in B.C. so all land is up for negotiation. Corporately, we feel it's fair – they're getting a fair shake," says McDonald.

"We have a good relationship with the Tahltan band – when I was CEO at Wheaton River, we built the gold heap leach mine in the same political riding as the Davidson Deposit – so, it's the same regulators and we have some history there. And, we have a rapport with the Tahltan band – for instance, we have over 40% employment by Tahltan First Nation at the mine. Overall, it's beneficial for everyone."

Thus reaching a goal of balancing financial rewards, environmental responsibility, and the interests of local communities. "Animals protected and prosperity galore in the town – that to me is what it's all about – a win-win-win situation," smiles McDonald.

Naturally, the recipe for this success includes having a great staff on board, with years of molybdenum and mining expertise; a critical part of ensuring the company's past and continuing success.

Kevin Loughrey is Thompson Creek's President, Chief Executive Officer and Director. Prior to the Thompson Creek acquisition by Blue Pearl, Loughrey was the President of Thompson Creek Metals Company and as such was responsible for all of the Thompson Creek operations. He negotiated, on behalf of the Thompson Creek shareholders, during the acquisition process.

"Like me, Kevin isn't a mining engineer either – he's a securities lawyer; better suited to the CEO title," McDonald says. Loughrey has also been the Senior Vice President and General Counsel for First Dynasty Mines Ltd. and Cyprus Minerals Company. Loughrey was instrumental in getting Cyprus listed on the New York Stock Exchange, and was prominent in the 1993 merger of Cyprus with Amax Metals Company. In total, Loughrey brings 27 years of mining experience to the table.

Kenneth W. Collison is the company's Chief Operating Officer. He has had more than 20 years of senior responsibility for mining operations and environmental affairs. Prior to joining Blue Pearl, Collison was Engineering Manager for the State of Alaska's Transportation Department, with particular emphasis on highway and marine terminal projects and Environmental Impact studies. He has also served as Vice-President and General Manager for Coeur d'Alene Mines (Alaska), Vice-President of

Crandon Mining Corp., a zinc-lead mine in Wisconsin owned by Rio Algom Limited and Exxon Coal and Minerals, and President of Rio Kemptville Tin Company, Rio Algom's East Kemptville tin project in Nova Scotia. He holds a B.Sc. in Mining Engineering and a Masters of Engineering in Mining from the University of Saskatchewan.

Peter N. Tredger is Vice President, Special Projects. He has been a senior officer of Blue Pearl since 2004. A professional engineer with 35 years of mining industry experience, Tredger has held senior management positions with Glencairn from 2002 to 2004, and Wheaton River Minerals Ltd. from 1992 to 2001. Previously, Mr. Tredger was an independent mining consultant, and for 11 years was employed by Amax Inc. in a variety of technical and management positions.

"Ken Collison, Peter Tredger and myself were the only three full-time employees in Blue Pearl before we purchased Thompson Creek – we were the guys," notes McDonald. "But, also I've had a partner in the mining business for 25 years – Kerry Knoll."

Kerry J. Knoll is a Director at Thompson Creek and is currently the Chairman of the Board of Glencairn. Prior to his appointment as Chairman, he was President of Glencairn since its incorporation in 1987. He was also the co-founder of Wheaton River and stayed with that company for 11 years in various capacities, including President. A graduate of Ryerson, Knoll has a background in journalism and has worked as the editor of both The Northern Miner Magazine and the Canadian Mining Journal.

"John Kalmet is officially retiring, not standing for re-election, but he's terrific – he's been my mentor for more than 10 years in the mining business. He retired when we sold Wheaton River," states McDonald.

Kalmet, a Director, is a professional engineer by trade. He was President and Chief Operating Officer of Wheaton River Minerals Ltd from 1996 to 2002. Prior to this, from 1989 to 1992, Kalmet was Vice President, Operations of Canamax Resources Inc. and, from 1964 to 1989, he was General Manager, Western Canada, of Noranda Minerals Inc. Currently a director of Glencairn Gold Corporation and North American Tungsten Corporation Ltd. he holds a Bachelor of Applied Science from the University of Toronto.

McDonald adds, "Jim Ashcroft has been here all through these negotiations, he was President of the Ontario Division of Inco Ltd." For 30 years, James W Ashcroft was with Canadian mining company Inco – the world's second largest producer of nickel – until his retirement in 1998. He is currently a Director with Thompson Creek, as well as President of J.W.

Ashcroft and Associates Ltd., a private company involved in mine safety and environmental planning.

Additionally, Mark Wilson is Vice President of Sales and Marketing. He joined Thompson Creek Metals Company in 2005. Having worked for more than 20 years in the mining industry, Wilson has extensive experience in marketing, business development and finance. Prior to joining Thompson Creek, he consulted for Climax Molybdenum Company on new product development (2001-2002) and served as President, Chief Executive Officer and Chief Financial Officer for Goldbelt Resources Ltd., a Canadian public company focused on mineral exploration in Kazakhstan (1996-1999).

From 1981 to 1996, Wilson was employed by Cyprus Amax Minerals Company in increasingly responsible roles including Vice President of Business Development and Manager of Molybdenum Marketing. Mr. Wilson holds a B.S. in Geology and Geophysics from Yale University and a M.A. in Law and Diplomacy from the Fletcher School of Law and Diplomacy.

"Plus, I would like to mention that we have two new guys coming on the board as Directors, as well as our existing – we have a very active board. Their names are Jim Geyer and Tim Haddon."

James P. Geyer is Senior Vice President of Gold Reserve Inc., a mineral development company. "Jim is a mining engineer; he was on my board at Wheaton River; going to rejoin this company now. Incidentally, his first 11 years out of school were with Climax – the Climax mine in Colorado was the biggest molybdenum mine in the world for sixty years."

Timothy J. Haddon is President and CEO of International Natural Resource Management Co., a private company involved in mining industry investment and consultation, and Chairman of Anatolia Minerals Development Limited, a mineral exploration company. McDonald notes, "Tim is also a mining engineer; he was President of Amax Gold; was with them 20 years.

"Both interesting guys that are going to add a lot to Thompson Creek's future."

A future that is bright, with opportunity and further rewards for Ian McDonald and his team. "I think we're doing fantastic. We're growing the company, developing our own projects, such as increasing the longevity of the mines, possibly increasing the size of the Endako mill and mine operation, and bringing Davidson on – that's all organic growth that we control. There's nothing there that we can't do, there's always some completion risk but, this is business as usual."

McDonald also expressed that he was "quite confident" that the future holds another transaction, but it would greatly depend on a number of factors. "I don't think our game is to buy all the other juniors and then have the problem of financing and building those mines," McDonald elaborates. "Our board, Kevin Loughrey our CEO, and myself – we all feel that we would be better off buying something that's already in production. Like I said before, one that's already built is worth more than two or three on the drawing board."

Although exploration can be necessary for producers, Thompson Creek doesn't usually work with explorers looking for moly – and the reason is simple. "To be truthful, there's no shortage of low grade open pitable molybdenum deposits in the world. What there's a shortage of is moly deposits in production."

For this reason, it's not uncommon for possible 'transactions' and acquisitions to emerge. "We get shown some very interesting situations; we have a size, we've got a $2 billion market cap. So, it's not too much of a leap of faith for someone to want to deal with us. And we've got all the staff. We've got people that can operate."

Ian McDonald says, "I believe that, one day, people will look at Thompson Creek as a business case. If you distill it down, it was all about timing. As you go through life, timing is such an important factor. We were in the right place at the right time and we had the ability to execute."

Execute they did, and the results speak for themselves. McDonald is pleased to say, "We're very uniquely positioned. There's no geopolitical risk with our assets – they're in Canada and the United States. You couldn't get a better place. Maybe a little tougher if you were trying to build a new mine in the U.S. now, but these are already built. We're well financed. We're fully integrated – meaning we have the roasting capacity – because you have to add about a buck a pound to anybody who doesn't have a roaster. Getting one of those permitted would be very tricky." Certainly, Thompson Creek has many positives on its side – and an aggressive strategy for taking this pure moly play to the next level.

"We want to be the 'go to' people on molybdenum. We're going to grow deliberately, and with some caution – we have a lot to lose. But we are going to grow. And I think we can be a little more aggressive than some of the bigger established mining companies. We're managing our future. Not that somebody wouldn't want to buy the company; that could happen. But we're not going to be caught resting on our laurels."

McDonald concludes, "Ultimately, we're going to stay in the molybdenum business – live by the sword, die by the sword. We think the future looks very good."

With Ian McDonald's dedication and leadership, Blue Pearl's business acumen, and Thompson Creek's globally renowned reputation for molybdenum mining; the future looks very good indeed.

TORCH RIVER RESOURCES

President and CEO of Torch River Resources, Dr. William Pfaffenberger, couldn't be happier with the progress of the company he has headed since April 2006. Torch River's premier property, the Red Bird molybdenum–copper–rhenium property, recently increased the total tonnage of its resource by 51% and its contained molybdenum by 40%. And with aggressive drilling programs planned through 2007 and 2008, expectations are high that the resource will continue to grow.

Red Bird consists of five mineral claims in west central British Columbia, 133 kilometres southwest of Burns Lake and 105 kilometres north of Bella Coola. The isolated 1836- hectare property is Torch River's main molybdenum project, according to Pfaffenberger, a retired University of Victoria Mathematics Professor. As a man with two passions in life—teaching and mining—Pfaffenberger jumped at the opportunity to concentrate 100% on developing a mining opportunity as President of Torch River Resources. Pfaffenberger is new to running a public company, though, he says, "I have a private mining company, called Fundamental Resources. It was formed a little over ten years ago, by a group of ten shareholders. We accumulate deals and mining properties."

One of the properties Fundamental Resources held an interest in was the Yorke Hardy molybdenum deposit, now known as the Davidson deposit. Fundamental held a 50% interest in the deposit when it was dealt to Thompson Creek. As Pfaffenberger explains, as part of the deal with Thompson Creek "Fundamental negotiated a net smelter return on the entire deposit. Don Davidson was my partner in that agreement. When it begins to produce, we will get a 2.75% NSR—which is substantial, because it's a big deposit."

"Now, we had one other molybdenum property and that was Red Bird. The person behind all this, in terms of gathering the properties together, was Andris Kikauka, who was my geologist at Fundamental. He knew where the major deposits were and kept track of what was going on. Andris called me in February of 2003 to tell me that Red Bird had become open ground. "I didn't even know what Red Bird was!" laughs Pfaffenberger. "I asked him and he described the history of the property. Within a week, he was in there staking it. So, Red Bird went into Fundamental." The team at

Fundamental Resources eventually spun Red Bird off into Red Bird Resources. "We partnered with a person in Alberta, on a fifty-fifty basis. That partner was responsible for finding a home for Red Bird. And the partner found Torch River."

This was Pfaffenberger's first encounter with the Canadian company. "At that time, I had nothing to do with Torch River. The option agreement put Red Bird into Torch River. Then, what happened over the next year was that I became familiar with the people who are Directors and management of Torch River." Pfaffenberger admits he had no intentions of taking the helm of the company. "It was the farthest thing from my mind. I was just a few years from retirement," explains the 64-year-old CEO. "But, the circumstances were such that the directors approached me last spring to become President."

Since Torch River's additions of Red Bird and President Bill Pfaffenberger, things couldn't be better for the company. "Well, at Red Bird, we have not drilled a hole yet where we have not been successful," smiles Pfaffenberger. "You know, sometimes the grades are lower—they're not all screamers—but, this property has a lot of mineralization. It's hard to find rocks that don't have something in them."

A molybdenum–copper–rhenium project, Red Bird's history dates back nearly half a century. "Initially, the property was drilled by Phelps Dodge in the 1960s," says Pfaffenberger, "and then there was an option agreement with Craigmont Mines in the late 1970s."

Exploratory work to date has shown that Red Bird's deposit is comprised of three zones of molybdenum concentration, referred to as the Main, Southeast, and Southwest zones. "Craigmont Mines had actually planned on putting the property into production in early 1980," explains Pfaffenberger,. "But, of course, the price of moly turned down and it just kind of disappeared from the radar screen. Then the ground came open, about 2003, and, after we staked the ground and made the deal with Torch River, the option was to get 100% of the property. That agreement was signed in the late spring of 2005. The moly prices were well on their way up by that time. The then-President of the company was the one who put together the option agreement, and then I became President about one year later."

Red Bird boasted a sizeable resource even prior to its recent increase and updated NI 43-101 report. At the time it was staked for Fundamental, Red Bird was thought to host an NI 43-101 implied metal content of 107.9 million pounds of molybdenum. Exploration undertaken since Torch River optioned the property has continued to increase that resource. "We're now into our third program," says Pfaffenberger. "We have ten people up at Red Bird right now. Just this last weekend I was up there—it's good to go and kick the tires now and then. It is really a very nice place to work, although it's in the middle of nowhere. We've got a good cook, good camp manager, and they've built some better buildings for us. The drillers have their own cabin with a TV, VCR, and a whole bunch of stuff," he laughs. "They work two shifts, so the drills are turning twenty-four hours a day.

"There was a program done in 2005, which re-assayed all of the data from the previous company. Then in the spring of 2006, an NI 43-101 report was done by Giroux Consultants (Gary Giroux also completed the NI report for Thompson Creek on the Davidson Property). That report identified an inferred resource of 75 million tonnes grading 0.065% molybdenum and 0.07% copper, at a 0.03% Mo cut-off. So, the contained molybdenum, if you just look at the gross amounts, is about 108 million pounds, based on the previous work. Then, last summer, we did a drill program of 6000 feet—there's a substantial increase in the amount of reserves, from just six holes."

Those newest results were released at the end of August 2007 and the updated report included the drill results from Torch's 2006 drill program on the Red Bird property. A further update is expected to include their 2007 drill program, once the results of that program have been received and analyzed. In addition, the 2007 NI 43-101 report has reclassified 43.3 million tonnes of the total resource from the inferred to the indicated category (0.064% Mo, 61.2 million pounds). Essentially, the total tonnage of the resource, at a 0.03% Mo cut-off, has increased by 51% and the contained molybdenum has increased by 40%. "It's actually interesting to watch that the market usually doesn't react to things anymore, unless they're actually in the official reports," notes Torch River CEO Dr Bill Pfaffenberger, who is thrilled with the official increase at Red Bird.

Meanwhile, 2007's program continues to roll at Red Bird. "Our drilling started at the beginning of July, and we've now drilled seven holes in a new area found by our geologist in 2006 that had never been drilled before. We're going to call it the Western Extension, and again we're pulling molybdenum in all the holes. Torch's plans for this current program are to drill approximately 9000 feet in total. Samples from the first hole, drilled to a depth of 1430 feet, and also from the next three holes, have been sent to an assayer and results will be released once they are received.

Focusing on upcoming drilling in the future, Pfaffenberger says, "So, the drill program for this summer was about 9000 feet. And the one we had last summer was 6000 feet. And if I get my druthers, the drill program we will do next year will be somewhere around 30,000 feet. We'd have intentions to basically drill off a fairly large part of Red Bird in 2008."

In addition to the moly and copper reserves, Red Bird also contains notable amounts of a rare, strategic metal called rhenium. Rhenium is one of the most expensive metals on the planet. "Thus, there's been plenty of interest because rhenium is a strategic metal, although it's in very small use and production," explains Pfaffenberger. Indeed rhenium's supply-demand balance is currently tight. The world demand is 41 tonnes per year, and that demand is met by primary production of only 35 to 37 tonnes per year. The balance is derived largely from recycled material. The demand is expected to continue to increase in the year's ahead, and will likely accelerate.

"Yes, true, the demand is very small, but the value's very high. All of the easy rhenium to find in the world is gone; there were other high grade, very small sources, but most of them have disappeared. There are some molybdenum mines—not all moly mines—that produce small amounts of rhenium as a by-product. We're lucky enough that we've got it," says Torch River's Pfaffenberger.

"When Red Bird was initially being explored, rhenium was not on anyone's mind. Rhenium is not one of the thirty-one elements that we used to test for. So when the property actually went in to Torch River, in that first program in 2005, we did some selective assaying in higher-grade molybdenum areas, and we found that it does carry rhenium. Since then, the interesting thing is that in all the samples we've assayed, the deposit always carries some rhenium. Some of it's relatively low, but, in the very high grade areas, rhenium grades as high as 2.5 g/t—which is significant."

As Pfaffenberger explains, "the real potential is that the rhenium goes along with the molybdenum concentrate, and then when they roast to convert the moly, they could recover the rhenium in the smoke stacks as well. You know, there's still a lot of work to be done in terms of metallurgical testing, to know how much we could recover. But, certainly, when this ore body gets to the development stage, definitely the roaster is going to have to be one that has that rhenium recovery as a part of its operation.

Torch River Resources is already carefully considering how best to mine the Red Bird molybdenum–copper–rhenium deposit. Recently, Bill

Pfaffenberger received valuable input from a few close, and experienced, colleagues. "When we went up to Red Bird this last weekend, I took up Don Davidson, who is the one the Davidson deposit is named after. I also took up a fellow by the name of Ellsworth Dickson, who's the senior editor and co-owner of Resource World Magazine. It was an interesting experience because they were actually giving me suggestions about how the deposit might be developed.

"Craigmont's plan for Red Bird was to have three open pits. They were just going to open pit the deposit for about 34 million tonnes. The grade of the ore they planned to mine was about 0.12% Molybdenum, which would be world-class grade for an open pit moly mine," admits Pfaffenberger. "But, you know, it's the tonnage, in today's situation, that could pose a problem."

Along with today's high moly prices, comes the high cost of developing infrastructure. Pfaffenberger worries that 34 million tonnes of high grade ore might not be enough to offset the infrastructure costs of open-pit mining in a remote location like Red Bird.

"But," Pfaffenberger says "the suggestion Don Davidson made was interesting. He worked at the Climax mine, down in Colorado, and they were the first ones to use a mining method called Block Caving. What they did was put adits (tunnels) under the deposit and then basically bring the ore down by gravity and take it out. It's extremely cost effective. Both Don Davidson and Ellsworth Dickson were saying that that scenario should be looked at by some mining engineers. Because the Red Bird deposit sits up on a mountain well above the base level of the lake. if that method worked, it would be a much better way to develop than an open pit, which unfortunately leaves a fairly big footprint. This would basically give us the possibility of an imprint which would be extremely small. And you'd just truck the ore out from underground, then have it processed and the concentrate shipped off."

As Pfaffenberger notes, Red Bird is certainly remote. The mineral claims are bordered on two sides by Tweedsmuir Provincial Park, however the entire claim area is open to mineral exploration and development.. The property itself is open to the southwest and lies 45 kilometres from tidewater. We've also staked to the south, so we've staked a good portion of the route that would be the route out, so that we wouldn't have any problems. It's true that the mineral claims are not the main thing for putting in roads, but if you've got somebody who wants to be a pain in the butt, right, they can stake the mineral claims on something you're going to put a road through, and then you have to co-operate with them to bring minerals out. So, we've looked into the future and basically tried to get this so that it's a stand-alone project."

Access to Red Bird is via float plane in June through October, and by helicopter in winter months. Float plane access is available from Nimpo Lake and Bella Coola, located south of the deposit, and from Burns Lake or Houston to the northeast. Although the isolation poses some infrastructure challenges, it doesn't overly concern Pfaffenberger. In fact, he recognizes the innate benefits of Red Bird's location. "My own feeling is that it's a blessing in disguise," he says.

"I must admit Red Bird is about as isolated as I've seen, except for maybe going to the North West Territories. We have a park boundary to the north, and we have a park boundary to the east. And the property's open to the southwest for the possibility of a road that would take the concentrate out to tidewater. But, the blessing of it is that the property is so far from any civilization that there really wouldn't be any disturbance that would be a concern to your neighbours," he notes. "With what goes on today, that's a big thing. If you're close to a populous, they're fairly concerned about exactly what's going to be going on in your mining development."

In regards to any possible concerns from First Nations communities, Pfaffenberger says: "Well, you know, as with all of B.C., there are claims. The First Nations community that is principally concerned with this development would be the Bella Coola. The town of Bella Coola is about 100 kilometres away. Red Bird is really at the top end of their territory to the north.

"There's logging probably within about 15 or 20 kilometres, but the ground is right at the top of a pretty rugged valley, and there's never been anything developed there in the past. Electricity-wise, harnessing an affordable, green energy source is an option. There's a real possibility," says Pfaffenberger. "Red Bird is at altitude. The lake sits at probably a little over 3,000 feet. There's a fair water drop in the drainage that goes to the southwest. And, you know, they now have things that run off river power sources. It may not supply all the power, but it would certainly be possible to partner with somebody, with Torch River being the buyers, to actually put that kind of a facility in – and it's cheap power. As a back-up, or to supplement the possible river power source, there's the possibility, of course, of bringing in power from a regular electricity source about 60 kilometres away."

Assembling the details for Red Bird's future, Pfaffenberger seems astutely aware of the potential for molybdenum. And Torch River is well-situated to capitalize on this—if they play their cards correctly. "The reason we're trying to prove up substantially more reserves is that I think it's going to be fairly difficult to secure financing for a lot of future developments. I would expect that's the way the molybdenum market is going to go. Look at how difficult it was for Thompson Creek to raise $550 million last fall, when they were buying an asset that the year before had produced $400 million, after tax." The bigger the deposit, as Pfaffenberger sees it, the better your chances of getting it financed.

"Now, if you look at the scenario, my guess is since there is a shortage of supply of molybdenum now, and nothing of substance coming into production in the next three or four years; I think the moly story is going to take on a life of its own. And I expect the molybdenum price to go a lot higher." He adds, "But, I think when it will really get known by the public is when there is an actual shortage of supply. Molybdenum is not warehoused; it's sold directly from the producer to the user. There is no pipeline. So, if there's a shortage of supply, what will happen is that the users—whether they're using it for pipes, or as material for nuclear power plants, or in stainless and alloy steels—are going to have to delay projects because of the lack of this element. And that will carry its own dynamic."

Pfaffenberger also points out that "the larger world-class mining companies are not looking for molybdenum deposits. Period. They're not even looking for people who have molybdenum." Joint venture partners haven't been burning a path to Torch River's door, he says. "And one of the reasons is that most of them are looking for copper—copper being the king in terms of a world-wide production, somewhere around 44 billion pounds a year. Copper's the main game. And the big players have been swallowing up other big companies that produce a lot of copper. But at some point there will be interest in Red Bird. It may not be a player like Rio Tinto, or the other big mining companies, but when you look at medium-sized mining companies that actually have production and cash flows, projects like Red Bird I think will come onto their plate, and there's some possibilities of taking on a partnership."

He concludes, "So, that's what I'd be looking for: we'd develop an attractive enough looking deposit that at some point we'd partner with a more senior mining company. And they'd bring in the resources for the development. However, at this point, I'm focused on satisfying our option agreement: 2008 is our year. We've got a lot of things to do in 2008, and once we complete those, then we'll own 100% of the Red Bird project."

Furthermore, "We have filed work against all the claims, so we have them until 2016. And the maximum that you can take them forward is 10 years."

Red Bird isn't Torch River's only project in development, and molybdenum, copper, and rhenium aren't the only minerals in their inventory. Torch River's portfolio includes the Grouse Mountain silver prospect near Houston, B.C., and the High Rock Island Lake and Climpy gold properties in northeast Manitoba.

"Torch River inherited a gold prospect in Manitoba—a high-grade gold vein. It's called High Rock and it's good until 2012. Looking forward, at some point, we may get a hot precious metals market, then that'll be pretty interesting." The High Rock Island Lake property consists of nine mining claims covering 1301 hectares and the Climpy property consists of one claim covering 243 hectares. Torch River holds a 100% interest in all of the claims.

Since 1985, there has been prospecting, line cutting, surface trenching and sampling, diamond drilling, and pilot plant test milling on the Manitoba properties by the companies which previously held interest in the claims. Torch River had conducted an exploration program consisting of reclamation work and bulk sampling from 1997 to 2000. In 2000, the company completed a ten-hole drilling program on the Climpy property.

Torch River also took on the Grouse Mountain silver property in the summer of 2007. "It's polymetallic in some parts, it has some zinc as well," says Pfaffenberger. "We will do enough work on the property—initial exploration stage sampling and some geophysics—to keep hold of the property for the next couple of years."

Grouse Mountain has widespread precious and base metal mineralization grading to 4% copper, 312 oz/tonne silver, and 0.33 oz/tonne gold. The mineralization occurs in a number of veins. As presently defined, the veins are not economic because they are deemed too narrow. But their depth potential, and the potential that the vein mineralization may be part of a larger mineralization event, is unevaluated. Grouse Mountain's Julia Vein has a small adit, from which a 2.72 tonne sample was taken in approximately 1939, which produced 12,548 grams of silver and 85.3 kilograms of copper.

"Both High Rock and Grouse Mountain are very prospective properties," Pfaffenberger says. "In other words, you have to put the work in. But if interest was there, you know, they could develop into something."

Certainly Red Bird's prospects alone are sure to keep Torch River's President from a retirement well-earned by 38 years of teaching mathematics at the University of Victoria. Though the mandatory retirement age has disappeared for the universities in B.C., he really had no intention of staying beyond the normal retirement date.

"But, I actually retired a little bit early, because I took on the Presidency at Torch River. When that was offered to me, which was a surprise, I was President of the Faculty Association. I was going to be President of the Faculty Association until my normal retirement, so I had to unwind a whole bunch of issues, to be able to get out the door," he laughs.

Born in the U.S.A., with a PhD from the University of Oregon, Pfaffenberger joined the University of Victoria's faculty in 1969. "I was raised in California and I lived in the Northwest, and went to school in Oregon. I became a Canadian citizen in 1980 and, certainly, I'm a true blue Canadian. And Canada is a fascinating place to work."

Swapping one President's hat for another, and leaving behind his academic career, Pfaffenberger's passion for geology and mining has gone from sideline to full-time job. In addition to being the head of Torch River, Pfaffenberger still runs his private mining company. And his interest in mining actually stems back to the 1980s: "Somebody introduced me to it. I used to invest a lot in oil stocks. And someone said, 'You know, being in Canada, take a look at an annual report of a mining company.' I was absolutely fascinated. The mining companies that existed then were so interested in their projects and the people who lived in their towns. They took a really paternalistic view. You don't see companies doing that so much now. Plus, of course, the resources are finite, so they worried about what's going to happen when these things eventually close down and so on. So that fascination started, and I started investing in small mining companies, and then I got on the boards of a couple of small mining, Venture Exchange companies."

But despite his decades long interest, Pfaffenberger admits "I'm not the expert. I use experts in geology." But Pfaffenberger brings his own wealth of skills to Torch River, skills he developed as a professor. "Mainly just the skills of analysis and decision making," he explains. "I won the top teaching award at UVic one year. In fact, that was a relatively big thing in my career. I really enjoy research and I really enjoy teaching."

When Torch River offered Pfaffenberger the role of President, he admits that "a key part of that decision was their Chief Financial Officer in Calgary, a fellow by the name of Barry Pearson. He has a substantial financial background. Pearson works pretty much as a consultant now, but he has worked for really large companies, as a CMA. So, when they approached me, I first thought, 'There's no way I'm going to do this.' But, I told them I would look into it. When I talked to Barry, I said, 'In reality, with your background and my background, the two of us, we could run this company.' And Barry agreed."

With Torch's head office in Calgary, Alberta, Bill Pfaffenberger works out of his home office in Victoria. He notes, "The way that we work as partners is that, whereas normally the President of the company would get substantial options; Pearson and I split all the options, fifty-fifty. The way Torch River handles options is two thirds of the options go to Barry and I, and the other third is divided between the other three directors. So, that's our incentive—and it's worked very well."

He regards his role at Torch River as his swan song, adding that, "in jumping into being President of the company, I looked at it as a culmination of my interest in all the things that go on at the various levels of mining development. I could have just stood to the side and been my usual investor self, but the hands-on part is very interesting.

As Red Bird develops, things are likely to get even more interesting for Torch River's new President. The dimensions of the intrusive at Red Bird are 1200 metres by 1500 metres. "When you consider something of that size," says Pfaffenberger, "I think, when we get our report in, Red Bird will go to something like 120 million tonnes. And this thing could be a lot bigger than that. So, the drilling next summer, that'll be a fairly aggressive program." With plans to drill off a fairly large portion of the Red Bird deposit by the end of 2008, and with three other properties waiting in the wings, retirement is now the furthest thing from Pfaffenberger's mind.

WESTERN TROY CAPITAL RESOURCES

We've all heard of Helen of Troy. Some day, many of us may be familiar with Rex of Troy. The Rex is Rex Loesby, the professional mining engineer with over three decades of experience who is the President of Western Troy Capital Resources, a minerals exploration company with huge potential. The name of the firm does not truly capture the essence of Western Troy's exciting properties: a high grade molybdenum/copper project, 500 miles as the crow flies north of Montreal in Northern Quebec, and a very promising gold/copper prospect in Panama, both of which are more Northern and Southern than they are Western.

Rex Loesby flew into Toronto in the early summer of 2007 to meet with his Chairman of the Board, Stanley Mourin. The latter of the two principals of Western Troy is a charming, grey-haired former stockbroker of 73, who founded Western Troy, while financing a large number of junior exploration companies in the Canadian mining scene over the past 35 years. Loesby approaches Mourin's stylish home, located but a few blocks from the heart of Toronto's Bloor/University/Royal Ontario Museum corner, on a bicycle, an intriguing way to travel for a 57-year-old engineer from Colorado, who was born just outside of Seattle, Washington. But then, Rex Loesby is an intriguing man, with the background and drive that cries out for the kind of giant success which Western Troy may soon experience.

True, the famous Hemlo gold strike of the 1970s was discovered smack on the edge of the Trans-Canada Highway in central Ontario, which made that monster find even more attractive to its supporters, whereas the molybdenum and copper in Western Troy's MacLeod Lake project will require a new road to make it a mine. But, it took 76 drillings costing many millions to finally discover the rich veins of gold in Hemlo, while the $2-plus billion dollars worth of molybdenum and copper sitting in Northern Quebec was so exposed, it almost advertised itself to its present owners.

"Back in 1988 my geologist was flying around MacLeod Lake in a helicopter looking for uranium," reminisces Chairman Mourin, "and he noticed a rusty spot on the ground. He sat the chopper down on the spot and took a few chip samples off the boulders and thought to himself, 'Wow—copper and moly! This isn't supposed to be here!'" But it was. And so the MacLeod Lake Project was born.

Western Troy has a small office in Toronto. Rex Loesby breezes into Chairman Mourin's charming house and sits down on a sofa in the den. He has a welcoming face, a deep voice, and is wearing a buttoned-down shirt and khaki pants; perfect for biking. His sleeves are rolled-up, apt for a man who has spent his 32-year career discovering, financing, and climbing into and out of mines around the world. This man is no Bay Street/Wall Street type who plows money into mining projects that he neither understands nor visits. And he is as articulate as he is bright.

Rex was born in Kirkland, Washington in 1950 to a father who came to Arizona where a mining engineer took him under his wing. Loesby Senior was nearly accepted to West Point, but managed to put himself through chemical engineering school at the University of Colorado. This was during the Great Depression, and in his proud son's words, "did very well for himself and his family coming from where he had begun." He influenced his son, Rex, who was hiking through the North Cascade Mountains of Washington State, when he was only 12, and loved to run across "old mine workings; "it was always interesting for me to see what the old timers accomplished." It wasn't by chance that when he started at the University of Washington in Seattle in 1969, he grew obsessed with mining. "I saw it as a basic industry that would have value for a long time, and that there would always be a need for basic materials to build our societies. I also loved the wilderness and mountains and a mining career would certainly take me to many remote areas of the world."

At this time, the environmental movement was very active at UW, and "the mining industry had been a bit of a whipping-boy of the environmentalists." Loesby took action: "I figured, if you're going to change something and make it better, you have a better chance of making a difference from the inside rather than just complaining from the outside." He went through their engineering school, received a mining engineering scholarship, "did very well," and after graduation from their mining school, he had to choose between working underground in coal in Kentucky, moving to Arizona to be a surveyor in copper mines, or going to graduate school in business. The latter won out, and during one summer, Rex worked at an aluminum refinery in Ferndale, Washington, half-owned by AMAX, one of the big mining firms of that era. He enjoyed that labor, which lead to writing his Master's thesis on Canadian-US tax law as it related to mining. As a result, he was recommended by the Chief Financial Officer there to the AMAX people in New York, who hired him right out of business school.

Rex Loesby's next few decades went no less straight than a Northern Quebec highway—lots of lakes and mountains to loop around. He had heard of molybdenum in university only in passing and it didn't have much significance for him. He only knew that moly's major use during the First World War was to harden the gun barrels. At that time the AMAX owned Climax Mine in Colorado was the biggest in the world, producing about half the world's supply of molybdenum. Loesby was hired by the Strategic Planning Division of the Climax Moly division, which was then in the middle of developing the Henderson Mine, a $500-million project in Colorado. "It was underground with a big, long tunnel, a huge operation. I was right in the middle of it all, and it was really a lot of fun." His job was to estimate the costs of production at Henderson, so that Climax could determine whether they were going to make any money at Henderson.

As he recalls those days, moly was selling for about $2.40 a pound, so he would report to his superiors. "You need to get about $3.60 to make it worthwhile." Since moly demand was strong at the time, Climax was able to edge their price up to $3.60 per pound over the following year. Henderson reached full production in 1980, and although Loesby was there for only a couple of years, he still enthuses, "It was a great experience for me. I sure learned a lot."

Rex learned early and quickly that the mining industry is a very capital-intensive business, and it's extremely difficult to be an entrepreneur and do things on your own. "You've got to continually find ways to raise money." Loesby worked for about eight years for large companies, and finally realized that this was not where he wanted to spend the rest of his life. "I looked at the people who had labored for these huge firms for several decades and didn't like what I saw: guys were trapped in do-nothing jobs, working for gigantic mining enterprises that were always making guarantees that they couldn't follow through on." In 1982, he was married, his bride had a good job, and within six months he quit his latest position.

Providentially, after kicking around for a while looking for ideas, he recognized that in order to raise money, one must understand investors, as well as mines. He was still living in the Denver area, having moved to Colorado as a result of his work for Climax, so he briefly considered becoming a mining analyst for a Wall Street firm or a bank before realizing, "You should be in the trenches of an industry and deal directly with investors to really understand why they invest." He went into the brokerage business as a stockbroker (for Smith Barney), "just a grunt broker; no specialty," and whenever he saw a mining deal that he thought showed promise, he had more to say to his clients than he might for a "Boeing" or

"General Electric." This was "pure cold-calling, getting on the phone and talking to as many folks as I could, but I couldn't restrict myself to mining stocks; I had to find ideas that met the customer's needs."

Over four years, 1984 to 1988, he followed this path. In addition to the broker job, Rex taught investments to seniors and graduate business students at the University of Colorado at Denver. He recalls, "On Black Monday, October 19, 1987, after the market had crashed 600 points, I had to go teach the 3-hour class and act like I really understood what had happened." He had wisely kept in contact with colleagues with whom he had worked while with the mining companies. One day, he had lunch with a couple of friends who had visited an historic marble quarry near Aspen, Colorado, that had operated early in the 20th century, producing the stone for the Lincoln Memorial and the Tomb of the Unknowns in Washington, D.C. It was a very classy white marble," but the quarry had closed just prior to World War II. Rex's friends wanted to see how the three of them might get it going again. Neither of his partners had much mining expertise, one being a petroleum land manager and the other an attorney, so Loesby happily visited and studied the quarry, phoned up the Marble Institute of America, and then contacted the people who owned the once-great property. Even though the demand for marble had increased by 850% over the previous decade (shades of moly in the early 21st century), the owners had no interest in the business. So Rex and his partners got a lease on the quarry and began the process of getting the permits and raising the money to re-open the quarry.

Then tragedy struck. One of Loesby's partners was killed in a car wreck about a year into this process. The other partner then insisted on being President of the venture as he had the greater ownership position. Rex insisted on running the quarry because "I'm a mining guy!" The young entrepreneurs eventually raised $5 million through an English bank, and in the fall of 1990, the quarry re-opened. However, within just a few months, Loesby saw that his partner was out of his element running a mining company and was taking the venture in the wrong direction. Rex left the partnership, and he started to wonder whether he might help other people who have mineral properties to find money.

He worked for eight years as a mineral property broker, and, as he humorously describes it almost in shock, "Believe it or not, I did make a living." He looked at over 1500 projects around the world, from Alaska to Argentina, to Australia, Papua New Guinea, and beyond. Maybe one in ten were worthy of being shown to anyone, and only about one-tenth of those would sell. He took a commission only when a property sold. Of course, the commissions on mining properties which did sell "kept us alive," Loesby smiles, "we didn't get rich, but I met hundreds of interesting folks in the business and I got pretty good at looking at a project and quickly determining whether it had merit. I certainly saw a lot of scam deals and I developed a good feel for what's out there. The work provided the background to allow me to quickly judge what projects have a chance to make it."

There were still more learning experiences to come before Rex met Stanley Mourin of Western Troy. In the late 1990s, when the Bre-X fiasco hit the markets, "I was in the final stages of helping raise $50 million for a gold/copper company when Bre-X hit, causing everyone to back out. The market went south in a hurry." At about the same time, the backers of the marble quarry in Colorado let it fall into bankruptcy after it had posted years of losses. The bank in England phoned Loesby and asked what they could do with this heavy marble weight around its neck. There were major lawsuits on the horizon, people were stealing both equipment and marble literally by the ton, and Rex admitted to the bankers that he probably could not raise any money because the venture had failed under three different management teams. The best he could do was to try to salvage it and see if he could make it work. "It's all yours if you will just pay for the transaction costs" said the bank. Rex Loesby fought the two lawsuits and won one of them. In the other, a neighboring landowner had sued over rocks which had rolled down from the quarry onto his precious property. In a Norma Mae, Hollywood-like victory, the witty judge awarded the couple $1 for each rock, for a total of $4, although Loesby had to absorb the cost of fighting the suit.

The quarry is known as the Colorado Yule Marble Quarry and has well in excess of 600 years of supply in its bowels. Rex re-permitted and then operated the mine profitably from 1999 until mid-2003. "It was a great experience to be able to own and operate a mine, something that is pretty rare in this capital intensive business. Of course you learn a lot when you have to pay all the expenses and meet payroll out of your own pocket, comply with all the environmental and safety regulations, as well as supervise production, engineering, and find customers for every block of marble produced." Most of the marble was sold to the US Government for use in the National Cemetery system. But in 2003, the bidding rules were changed such that the Colorado Quarry was cut out of the business. In early 2004, Rex was able to find a buyer that had the financial strength to weather the difficult times and the quarry is still in production. "I'm mining engineer," he grins, "but I guess you can call me an entrepreneur, too."

After selling the marble quarry, Loesby began looking about for other projects. In late 2005 Rex was introduced to Stanley Mourin, born almost two decades before Rex. "I've got this molybdenum/copper project in a company called Western Troy. We need to raise some money to test more targets on the MacLeod Lake molybdenum/copper property," Mourin told Loesby. With his experience in molybdenum, Loesby immediately saw the unrecognized value in the project. "Stanley, this is a damned good asset!"

At age 17, Mourin had started out in 1954 as a messenger boy for brokers in Toronto at $35 a week. A broker gave him a try at selling stock over the phone and Stanley proved to be a natural born salesman. He worked is way up through the system and eventually owned his own broker/dealer firm, underwriting dozens of junior mining companies over the years. The MacLeod Lake moly/copper project is one that Stanley held on to. "I want it to be my legacy that we make a mine at MacLeod Lake," he says. To Mourin's credit, during the 1990's when molybdenum and copper prices were in the tank he persevered, and added potentially exciting resources as they kept the drilling and geology moving along.

The meeting of Rex and Stanley was fortuitous. Rex had the technical and management skills and experience to operate the company, and Stanley had Toronto street smarts. Rex became President of Western Troy in January of 2006. They had a half-million in their treasury, and were listed on the Toronto Venture Exchange, trading at 20 cents a share. When the public saw that Western Troy was on the move again, the stock tripled in price and Rex was able to do a $1.35 million offering to get more drilling done. As these things go, Western Troy drilled some targets in the spring and summer of 2006, but failed to get the results they expected, and even though the core asset was still intact, and the stock "ticked down to 20 cents again."

Then, in September of 2006, Rex was able to get the story of Western Troy to John A. Kaiser, the well-known mining newsletter journalist and publisher of the influential Kaiser Bottom-Fishing Report. Kaiser liked the project, loved the company, and recommended the stock. Its shares promptly doubled to 40 cents, and "suddenly everyone started to get interested in moly," chuckles Loesby. Kaiser wrote at the time, "Western Troy to benefit from the long awaited MolyMania. . . . The conditions are finally right for MolyMania to kick in and do justice to the juniors which have acquired molybdenum projects." From such commentaries, the share prices of small mining companies can skyrocket.

Other events affected Western Troy's stock. The price of molybdenum had exploded from $2.50 a pound in 1999 to $45 a pound by 2005, as stunning in its own way as was the gold price explosion of a few decades ago from the government-mandated $35-an-ounce price to over $800. When he hired on with Western Troy, Loesby recalls that "everyone was asking why get interested in moly? The price will only collapse back to the $2.50 range." But people have come around over the past year. Most now agree that such a drop will not happen, and that those companies which have really good projects, such as Western Troy, are worth far more than their stock prices reflected in 2006.

Eric Sprott, the widely-admired money manager and owner of Sprott Asset Management, got interested in Western Troy, and in March, 2007. As a result of this news, Western Troy's share price took off, reaching a new high of just over $2 a share, before settling back to the one dollar range.

WRY (on the TSX-V) is a thinly-traded stock. "Investors like the fact that we have so few shares outstanding," notes its President. This means, of course, that moves in the stock will often be exaggerated and attentive stock traders can play the swings. Still, the market cap of the company is "quite a bit less than it should be," and both he and Chairman Mourin believe it should grow substantially, based on the quality of the assets relative to other moly juniors out there. The latter's argument is that Western Troy has an advantage over most projects of its kind. First of all, the ore grade at MacLeod Lake is about 50% higher than the average of all operating or prospective surface moly mines. Second, half of the ore value is in moly and half in copper. "Most mines are either pure moly, or have a small amount of moly; we're about 50/50 in terms of gross metal value. If you have a pure moly project, you will be exposed to a lot more risk, because you're subject to only one price, while we get two prices – moly and copper – to work with."

As the great American novelist, Joseph Heller, once wrote, "there's always a catch—Catch-22." If there is any catch for Western Troy and its shareholders, it is one also filled with great promise. While its management believes that they have what looks to be a surface mineable resource, they are 140 kilometres from the nearest all weather road. So, it's all fly-in, and expensive to explore. But there is good news as well. As an incentive to mineral explorers in the Macleod Lake region, Quebec provides refunds of about 47 percent of the money spent on Quebec exploration projects. The Quebec government is also looking to improve a winter road near MacLeod Lake into an all-weather one, and the Cree First Nation is very supportive of the road upgrade and Western Troy's work to develop a mine at MacLeod Lake. The Cree are pressing the provincial government in Quebec City to finance the new road—and it's a government which is extremely eager to develop its mineral-rich north. Loesby launches into a powerful soliloquy worth quoting at length:

"If we increase the resource at MacLeod Lake with additional drilling, that should move the stock. Even if we don't expand the resource substantially, we currently have over 300 million pounds of copper and 44 million pounds of moly. With moly prices at about $30 per pound and copper prices at about $3.50 per pound in mid-2007, you can do the math. Out of the 40 or so potential new moly or copper/moly projects that have defined resources, the MacLeod Lake resource is the fourth highest grade. After this summer's drilling program, we plan to generate a scoping study and this will lead into a full feasibility study next year. We hope to get the commitment from the provincial government to build the road sometime over the next year and this could also move the stock."

Rex and Stanley of Western Troy have no desire to be one of the big producers, but rather to bite off just a bit of the market. After all, if a company starts producing 40 million pounds of moly in a 400 million pound market, it had better be ready to cut back production when prices start to weaken. Loesby recalls the Henderson Mine with some regret: "Think of all those years between 1983 and 1999 when they produced 20 to 30 million pounds a year and sold it at $2.50 and $3.00 a pound. Think of the shareholder wealth lost as a result of this. If they had just put that mine on care and maintenance for a few years and waited for the market to come back, they could have made much more money. Of course, hindsight is 20/20. My thought is that the big producers such as Freeport, Codelco, Grupo Mexico, and Kennecott, who together control over half the market, have learned that lesson."

The leaders of Western Troy insist that their modest size means they are not a threat to generate moly market surpluses. With an estimated initial capital requirement of $150 million, Loesby believes a mine can open and start producing at MacLeod Lake, although a full feasibility study will be needed before the actual cost can be more accurately estimated. Another advantage MacLeod Lake may have over other pure moly projects is that the expected copper production can be used to secure debt financing. This is because there is a copper futures market – something that does not currently exist for moly. When closer to production, Loesby hopes to have a strong stock price, to raise the equity, get the mine open, and avoid diluting shareholders more than necessary.

Like many others, Loesby and Mourin do not see the recent gigantic leap in moly prices as a real mania. Look around us, they note. "The Chinese economy has been growing for 20 years at over 10% a year with no sign of slowing, and India has a billion-plus people in an economy that is in the early stages of a similar growth pattern. Brazil, whose economy is certainly not small, is growing at over 5% per year," Rex adds, and "Even Germany that has been stagnant for a number of years saw 3.5% expansion last year." All that steel in all those new buildings is going to need molybdenum, and few today would argue that simple fact.

Adds Mourin,"I just read an article describing the planned construction over the next ten years of over 81,000 miles of oil and natural gas pipelines, using molybdenum to strengthen every inch, especially in Russia and the Canadian North. Moly is uniquely suited for toughening steel in cold weather extremes. Nuclear plant pipes need moly, so with the worldwide boom in nuclear plant construction, demand from this sector is expected to grow dramatically." Loesby, in an insightful article he wrote for a mining engineering magazine in the summer of 2007 entitled Molybdenum Supply Forecasting ("because part of my job is to educate people about the mineral") makes a strong case that moly prices will not drop below $15 for the foreseeable future, and will easily stay in the $15-$30 range.

The world has gone through 20 to 30 years of difficult times in terms of opening new mines, partly a result of the new environmental restrictions and sometimes irrational public resistance to developing mines. Both Loesby and Mourin stuck with it during those years, when people didn't want to get into mining because they thought it was a depressed and nasty business. It was hard to stay in the industry, there were few jobs, and most commodity prices were as depressed as that of molybdenum. Rex shakes his head when he notes that of those with whom he attended mining school—both geologists and mining engineers—he is the only one whom he believes has stayed in the business. "In the mining and geology fields, there are now a few inexperienced people in their 20's and 30's, and lots of experienced people in their 60's and 70's, but not too many folks in between."

For Western Troy in northern Quebec, Loesby sees a 10- to 12-year lifespan for the project with the existing resource, but he believes there are more resources in the area. He thinks that because of the geology of the area, a good geologist could spend 20 years or more finding more ore for the mine. So the mine's life could be much longer. An environmental assessment is complete, and a final feasibility study is needed before they can submit permit applications and raise project financing. He feels that 2010 or 2011 would be a reasonable timeframe for production. What reassures both Loesby and Mourin is that they've now got plenty of money in the treasury, making it unnecessary to pound on doors of major companies for a partner, or dilute the stock in the near future.

Most companies want some back-up to their growing plans, growing pains and growing fortunes, and Western Troy's management believes they have found it in Panama. And it's not only its weather which is slightly better than northern Quebec: "Our project there has huge potential," insists Rex Loesby. "It is not going to cost a lot to find out if we've got something really big. We have experienced people doing the work there and it didn't cost much to secure an option to earn 70% of the property." They're currently exploring an area near the hamlet of Viento Frio along the country's north coast, two hours from its capital, Panama City, and just east of the justly-famous canal. Stan Mourin describes the project as having good promise for finding gold, copper and molybdenum.

The men are proud of their board: "The people who have been involved with the company from the start are professional mining people, well-know and well respected," asserts Loesby. One board member is Steve Vaughan, a mining/securities law specialist with over a third-of-a-century of experience; "the most prominent mining lawyer in Canada," adds Mourin. Then, there are Consulting Geologist, Stewart Winter, who has nearly a half-century of Canadian mining industry experience; Wayne Holmstead, their Vice President of Exploration, with over three decades in the field; and Director Edward Thompson, another mining engineer coming close to the half-century mark with his own work in mineral exploration and junior mining management. Their mid-town Toronto office is not large, and with their drilling contractors and helpers, there are about a dozen people on site during drilling programs.

When asked what drives him in this business, Rex Loesby has a surprisingly personal response: "I have two daughters in private universities in the US and a son who is looking at Stanford or Cal Tech. So that means high tuition expenses over the next many years. That's a major motivator." But next to a wonderful family, the President is motivated by the ambition to succeed with integrity. "I have seen many people over the years compromise their integrity to make a buck. I believe the real challenge in life is to make it without making that compromise."

Loesby says he is "very happy with where Western Troy is with our properties today. We're still in a bit of shock. In January of 2007, we had only a half-million in our treasury, and wanted to do a million dollars worth of work this year. We wondered how we could do that with a stock price of 40 cents. Then, we raised $4.5 million over the next two months. First a half-million came to us through MineralFields in Toronto. Then Eric Sprott and Pinetree Capital came along and offered us the rest. With that money, we are now able to do far more work than we hoped for, to advance more quickly at MacLeod Lake, as well as add projects like Panama to the Western Troy mix of exploration and development properties."

Let Stanley Mourin give his own prediction of the future of Western Troy, the company he nurtured for nearly two full decades: "I've always felt that this could be a very valuable stock, maybe even more. We staked the Crown land, we worked it, and we're finally getting it to the point where it's proving to be a wonderful project. It's real, not just hopes and dreams, and we're confident we can increase the resource. I want to see it become a mine."

WIN-ELDRICH MINES LTD.

A historically illustrious company and long-time member of the TSX-V, Win-Eldrich Mines Ltd. has remained under the wing of 84-year-old CEO Reuben Brant for 20 years. After waiting for a rise in metal prices, finally the climate is ideal for their Nevada-based molybdenum and gold property, Ashdown.

"It turned out good with the moly," says Brant with a wry smile. "It's got the company going; the stock has moved from 15 cents to 3 dollars. I've always liked gold, but moly is one of the real metals that is in demand with pipelines and everything – it just keeps on going." A bit like Reuben Brant himself - who has suffered in recent years from some health problems, yet keeps going. "Well, you gotta keep going, what can you say¿"

Starting off with an eye on the low-grade gold, which is present at Ashdown, the high-grade molybdenum mine now features prominently in Win-Eldrich's future plans for the project. "It took 20 years to get the darn thing going," notes Brant. "You couldn't do much during the days when gold was low."

Now looking toward the next two decades, and further, the company's President, Perry Muller, and fellow directors are quick to commend Reuben Brant's remarkable resilience and dedication, which has kept Win-Eldrich around long enough to benefit from today's prices. Indubitably, Brant's decision to pay out of his own pocket to hold onto the Ashdown property, in spite of dire market conditions, seems as astute as it was risky.

"When I took over Win-Eldrich, for a long time I was putting up my own money – they never expected moly to move," says Brant, who still actively participates in decision making and board meetings, usually in person, sometimes via teleconference from his home in Toronto. "If we go back to when we made the deal on the property, we got some financing from a good friend of mine – Ed Gaylord. He put up quite a bit of money to keep things going and paid taxes, too; they build up in a hurry. He owned newspapers, hotels and the Grand Ole Opry in Nashville," and his family now runs his multibillion dollar empire that includes radio, cable and TV stations. "He's partly responsible for all this going on," says Brant.

"Ed looked at the Ashdown property (located in Humboldt County, Nevada) and said there's no reason why this isn't a big deposit. He put up money at a time when it wasn't easy to raise money. His family still owns about 3 million shares in the company," says Brant.

"Meanwhile, today, we're going through a very different market. Things are really happening; you can raise financing. But, I got Ashdown for next to nothing." Thus, as a result of Brant's diligent payments, the company can proudly state that they have absolutely no debt.

Not bad considering that "Win-Eldrich is one of the oldest listed companies on the Toronto stock exchange," as observed by William Sheridan, a company director and legal counsel, who's experience predates even Reuben taking control of the mining firm. "This is a company with deep roots in the business; first started in 1942."

At approximately the same time, the Second World War began, and the U.S. government consequently closed their mines to enlist workers as soldiers. The Canadian mining industry, however, flourished and a company founded by two men, Win and Eldrich, decided to invest in mining claims in Ontario, beside the successful Lac claims. Nevertheless, Win-Eldrich left empty-handed on this occasion; the claims were mined out by the 1950s, and the company languished before being bought as a shell and reorganized several times, for oil and gas, and other minerals. This continued for 40 years, until Reuben Brant took an interest and revitalized Win-Eldrich Mines Ltd.

"Certainly, Win-Eldrich is old – dating back to its incorporation in the 1940s – and it's something that's very rare because most of these companies went bust. They continually diluted their shares down to nothing, or it was a lot of smoke and mirrors," says Perry Muller – a shareholder and businessman so taken with Reuben Brant's hearty efforts that he became involved as a Director and then company President.

"Win-Eldrich acquired an interest in Ashdown in 1987," says Sheridan. "They paid off all underlying property owners and royalty holders, then in 2004, Win-Eldrich entered into a 40/60 joint-venture with a mining company in Nevada, Golden Phoenix Minerals." An agreement was signed between the two companies to create the Ashdown Project LLC.

Muller became involved the same year. "What we've done in the last few years, that would be my contribution to the company," he says. "Basically, in 2004, I was offered three stocks and I bought all three and they weren't going anywhere. So, I decided, well, maybe I should check out what kind of an investment I made. I called up Win-Eldrich and I got to speak with Reuben. We started talking and we struck up a relationship. He started calling me back, and next thing I know, I said 'okay' and I went to Toronto and offered my services."

Born and raised in Nebraska, where he owns United Transport Inc., a trucking company; 46-year-old Perry Muller noted that the Ashdown project was progressing slowly. "I decided to spend my money and my time to see if I could make this thing speed up. Because it's a great, money-making property and company; it just needs to be driven," explains Muller. "So, Reuben and I strike up this relationship, and I met with him and another director, Fern Ferreira (currently Vice President of Win-Eldrich), and I talk to some of the Golden Phoenix people, saying: 'hey, what's taking so long? What's the progress? What's the process we're going through to bring this thing along?'" In 2005, Muller took on a directorship at Win-Eldrich, and then assumed the role of President in 2006.

Prior to Golden Phoenix's involvement, "we made deals but never completed," notes Reuben Brant. "Billiton made a deal to bring our property into production, but the price of gold dropped, and molybdenum dropped to next to nothing. So, I did what I had to, to keep the company alive. But, I was always convinced that when you have an ore body, or some ore, that sooner or later you're going to find a bigger deposit, and that's what happened."

Originally from Medicine Hat, Alberta, Brant has literally been in the business of 'pulling something out of the ground' for most of his life. "That's right – that's where the oil and gas is, in Alberta. And, don't forget, I owned a brokerage firm for 15 years; Brant Securities," he rightly says.

"I actually came down to Toronto from the University of Saskatchewan with the intent of becoming a professor, but you have to have postgraduate training. So, I came and enrolled here and spent the money my dear mother gave me, and that's how I ended up in Toronto," he laughs as he reminisces. Along with over 50 years exposure to the mining world, which began with his association with Lakeshore Gold Mines (Lac Group of mining companies); Brant earned a B.A., B.Comm., and M.Comm. in Business Administration and Economics from Saskatchewan and Toronto universities. "It's been great here, you know," he says.

"Reuben's on top of his game. He's as sharp as ever," says President Perry Muller. "I want the awareness of Win-Eldrich to come out, as being this diamond in the rough, but I also want to pay tribute to Reuben."

"Reuben does deserve a lot of credit," agrees Bill Sheridan. "He put money in to keep the company going during the dark days, when nobody cared about small companies like Win-Eldrich. He truly kept the company alive.

"It's a tribute to Reuben that we only have 13 million shares outstanding and no debt – we don't have creditors breathing down our necks and we don't have a situation with 150 million shares outstanding. We've got a nice little company, a good group of shareholders, and we're all waiting to see what is going to pan out of Ashdown."

Sheridan concludes, "Reuben was the guiding light in the acquisition of this property, and ultimately the deal with Golden Phoenix, which has had its ups and downs, not as smooth as we'd like, but we're mining and selling molybdenum."

"There was roughly 2000 tonnes of product that was sitting above ground; piles that sat there for 27 years. It was severed ore that came with the property, that needed to be processed. It was mined by, I believe, the American Copper & Nickel Company (ACNC, the U.S. subsidiary of INCO)," says Muller. In 1987, Win-Eldrich acquired a joint-venture interest in the project from ACNC, before buying out ACNC's remaining interest in the mid1990s.

Times have definitely changed, however. "Most of the period of our ownership, molybdenum was at $2 a pound, and I can remember no discussion of moly reserves on the property," states Win-Eldrich director Bill Sheridan. "What sparked enthusiasm in 2004 was the rising moly price. And it's staying up there – it's persisting."

"Molybdenum was originally discovered at Ashdown in the 1960s," elaborates Pete Winn, the geologist at Ashdown. "It was a mineral of interest, but not economic. In 1979, ACNC bought a controlling interest in Ashdown; looking at its gold. They spent several years doing that and failed due to the small size of the deposit - 100,000 ounces at most, heap-leachable – too small for huge companies, and there was the fact that gold prices collapsed."

Ashdown has a past that includes underground production of 16,000 ounces of gold, at an average grade of about 0.36 opt (ounces per tonne). Reportedly, free gold occurs in thick quartz veins with gold showings in all directions from the defined resource. These were initially drilled by ACNC in the mid-1980s. Eventually, follow-up drilling and feasibility studies were conducted by Win-Eldrich, Billiton (Royal Dutch Shell) and North Lily Mining in the late 1980s to early 1990s, resulting in proven reserves of about one million tonnes with an average recoverable grade of about 0.075 opt. The Ashdown Project LLC plans to redo the earlier feasibility studies to determine whether the gold resource is still economic.

Nevertheless, in total, ACNC drilled 34 surface and 14 subsurface holes in the Sylvia Vein from 1981 to 1983. The Sylvia Vein is structurally related to the Ashdown Gold Vein to the north of the property, and it is exposed in Vicksburg Canyon, where additional drilling by ACNC indicated more molybdenum mineralization – a total strike length of moly mineralization in the vein system is over one mile. Incredibly, grades as high as 15% Mo were encountered in drilling and underground sampling, among the highest moly grades ever reported.

Winn says, "In 1982, moly prices skyrocketed from $2.50 a pound to $18 a pound. INCO, being a nickel producer, and nickel being used in steel, took a second look at the moly there. American Copper & Nickel drove a decline in 1982 and 1983 about 1000 feet into a world class high-grade zone – around 3% moly." ACNC took 32 multi-tonne bulk samples.

"They developed this moly decline, and produced about 1,500 tonnes of 2.5% moly, and then another 500 tonnes of 0.2% moly in the surrounding zone. ACNC then crushed it and planned a bulk metallurgical test, but moly prices crashed. They left a stockpile there.

"I sold this project to Win-Eldrich in 1986, with the understanding that it was a pretty small gold deposit and moly prices were down in the dumps. I was a manager of this project til late 2003," notes Pete Winn. "After temporarily removing myself in 2005, I was asked back and I've been consulting for the company since then."

Winn explains, "Moly sulphide is what we produce, not moly metal or moly oxide. Moly is priced by moly oxide, which is produced by roasters. We sell our product to a broker who sells it to roasters.

Perry Muller says, "Usually, high grade veins have a huge porphyry somewhere. In historic mines that have been similar to this one, when you get high grade ore like this, you have a porphyry system; either it's molybdenum or copper."

In the end, Win-Eldrich was halted by undesirable prices. "We had subsequently doubled the gold reserves – but, those prices started to decline in 1994," says geologist Pete Winn. "So, we ended up sitting on it, and Reuben Brant just said, 'I think I'm going to hold it.' He put the money in every year for the holding costs – a big risk, but he stuck with it."

Fortunately for Win-Eldrich's present shareholders, Brant did stick with it, and they are able to capitalize on this valuable moly stockpile. "There's been some pretty good-sized players that have pumped a significant amount of money into this thing," says Perry Muller. "Now we've moved the moly to an independent mill in Montana, to make sure we could turn it into a product that's saleable. Because if we do that, then we know our mine's a home run. So, we raised the money and moved the stockpiles, and we found a mill that would process the stuff, and we turned it into a saleable product and sold it. We moved a lot of trucks to get processed. And we made money on it," says Muller, a man familiar with the trucking game.

Furthermore, in recent months, as a minority owner of Ashdown Project LLC's molybdenum mine, Win-Eldrich announced that, based on information provided and publicly disclosed by their joint-venture partner Golden Phoenix Minerals, Inc., the mine's manager and majority owner (after earning a 60% interest in the property in 2006, by spending $5 million on mine development) during the second quarter of 2007, the Ashdown Project LLC posted sales of molybdenum concentrates totaling US$4,466,395 The Ashdown Project LLC has projected $10 million in sales for 2007.

While expenditures over the past six months exceeded initial budgeted amounts, Golden

Phoenix says that as the project evolves from a focus on infrastructure development to concurrent mining of multiple 'stopes' - the working areas of the mine; an increase in the production and sale of concentrates, together with a decrease in the project's total cost per pound, is expected. Until then, ongoing development work and equipment upgrades and repairs were conducted during the third quarter of 2007, which is expected to reduce financial performance and the rate of expanded production in the third quarter, but should position the Ashdown project for sustainable full-production rates later in the year. They expect to reach a production capacity of 100 tonnes-per-day by the end of 2007.

"Yes, you get a lot of money for one of those truckloads. At present prices, moly becomes very commercial," points out CEO Reuben Brant. "So, there's lots of potential.

"Of course, we never saw markets like this," he continues. "If anybody had of known that you were going to come up with prices for metals like you have now, you wouldn't have believed it!"

Believe it or not, the Ashdown project seems well situated to 'cash in'. Therefore, the company is moving and selling moly, while weighing up their future options. "Golden Phoenix has done an IP survey, identifying an extension of the reserves - but we don't know as it hasn't been drilled," notes Pete Winn. "Our drilling doesn't match our underground results, which is not uncommon for a high-grade deposit. Yet, it presents a metallurgical challenge for reserve projections."

Essentially, "small companies need to spend their money focusing on what they've got; they can't spend a lot of money on exploration until you know what you have. Though, I think there's a lot of potential at Ashdown. Within the next year, we'll have the money to look at the nature of the project, nature of the reserve, the geology, etc.," says Winn.

President Perry Muller is optimistic that dollar signs are on the cards. "I think that little Win-Eldrich is going to take their Ashdown project and use that as a cash cow. There's a lot of value there. Now we can spend that many ways; one you can return it all to the shareholder, two, I think we will probably raise some more money, and we can work on some other properties."

This might potentially include Canadian properties, should the opportunity arise. "Oh yes. If there's something decent there, you have to take a chance," says Reuben Brant, speaking with a lifetime of experience as an owner of a Canadian securities firm and a mining company.

Muller states, "In fact, I foresee us raising some more money, if needed, and working on our Cowden property, which is a gold property; it's very high-grade."

Also in Nevada, just ten miles away from Ashdown, Win-Eldrich has an option to earn a 100% interest in Cowden. Previously, about 30,000 ounces of gold at an average grade of approximately 1.0 opt has been produced at Cowden. Underground sampling confirms higher grades, but recent drilling below existing workings, in the winter of 2006, was deemed inconclusive.

Although, due to Cowden's proximity to Ashdown, if gold was developed at both sites, their ore could possibly be shipped to the same leach facility – making Cowden a particularly interesting exploration target.

"We're mapping it out right now," says Muller. "We have a geologist looking at all the old workings. There's been a number of them - whether it's five or six thousand feet of workings. They took a lot of product out, but his job is to map it out and see where they went to take the gold out and then put it in a 3D form to give us good drill targets. Basically, figure out where this vein has actually gone. Because it hasn't been mined since before 1941, I believe."

A National Instrument compliant resource estimate will likely be on Win-Eldrich's agenda. "I believe we have to at some point. But, right now, we're too early on that. And I'm one of these guys – like Reuben – who believes that gold is going to go higher."

Regardless, Ashdown is a 'sure thing' as a producing asset, while Cowden is considered a developmental/exploration type of project, for the time being. "The way I see it," says Muller, "Cowden is a roll of the dice and Ashdown is the money in the bank. It's hard to roll the dice, obviously, but everyday gets a little clearer. And Ashdown has tremendous upside. Ashdown should take care of itself and spin out money for us. That's a no-brainer, I would say.

"We've already permitted, we're taking ore out of the ground, we're processing it through our mill, and we're selling product, even through it's only a 100-tonne-mill and it's small in size, it has large returns on revenue."

President Perry Muller – an interested shareholder who was inspired to get involved and do the footwork for Win-Eldrich, trying to speed up the process to put this thing into production – admits that he is motivated, like many shareholders, because "I just want my stock to go up." Muller is presently the largest individual shareholder at Win-Eldrich, but he says that becoming President of the company "is something that's come around. I didn't set it up to run Win-Eldrich – it's a stretch to think you're going to go from a shareholder to a director to the president, in a short period of time. So, it wasn't something I set out to do. Yet, I thoroughly enjoy doing it.

"I mean, this was my first mining company, per se. I come from the private sector, so this is also the first public company I've been involved in. Public companies are new to me, and they move a lot slower than private companies. So, when people look at this saying, 'you move quickly, you came in the fall of 2004, and you're a director in the Summer of 2005, and you're president in the Summer of 2006'; believe me, I don't think it moved fast at all. When your money's on the line, it doesn't seem like it moves fast enough.

"But, it's a bit of a 'love story' in that I came back to the States after learning more about Win-Eldrich and its properties, and I'm buying into this thing because I really believe that Reuben is very sincere. I believe that he's really got something."

"Something' that a larger company might be interested in, down the line. "We've seen some interest of maybe a takeover but nothing has materialized to date," notes Reuben Brant.

Perry Muller adds, "Would it upset me if I sold the company for $7 or $10 a share and made all the shareholders money and walked away? No, not at all. If the offer was on the table, and the price was right, I would sell Win-Eldrich and walk away. If it's in every shareholder's best interest, I would do it. So, that goes against my ambition of wanting to run this company forever, because that might be it. You've got to do what's right for everybody, and our obligation is to the shareholders.

Offers of takeovers or not, with a portfolio of two interesting properties with gold and molybdenum potential; further acquisitions are still a definite possibility for Win-Eldrich. "I think the future of the company is going to be wherever the next project is. Where the goods are, where the silver lining is – which is not just Nevada, it could be Canada," continues Muller. "There's so much hype in the mining industry, it's unbelievable. And if you've been around it, they spin everything. But, if the project is real, then I believe we should go there.

"Our objective is simple – make money, and hopefully pay dividends and also bring on new projects. It's that easy. We don't have to go out and do a project. I mean, our desire isn't to be Barrick Gold, or somebody of that nature.

"My desire is to make more – what's the best way to put this – from a per share basis; we want more value than any other shares in the mining business out there. So, we don't care how big you are, how small you are, we just want to bring more value to our shareholders, on a per share basis. Which means dividends, hopefully, which means new projects that have a tremendous upside, things of that nature. We don't want to just be a mining company. We want to be a profitable mining company that the shareholders can benefit from. We want shareholders to own Win-Eldrich; we don't really want shareholders to trade Win-Eldrich."

Naturally, Ashdown's high-grade moly is going to play a large part in their future plans, which Perry Muller equates with a 'survival of the fittest' rationale. "If I was predicting, as of today, I believe moly will go higher. But, I do also believe moly will have a slight downward trend in a few years. And then what'll happen is, it'll end up being the survival of the fittest; just like every other business, or any other commodity.

"We believe, since we have such high-grade moly, we should be the low cost producer - which should make us the fittest. It's just common sense, if I have a higher grade product, I'm going to make more money as long as I control my costs. If Ashdown Project LLC can control its costs, there's no reason why they shouldn't become the low cost producer out there."

Reuben Brant nods enthusiastically, "Oh yes, and we expect to see better things yet." The mining business still gets him excited after all these years; having celebrated his 84th birthday in October 2007. How did he celebrate? "Well, I had a little bit of a party just knowing that, I've always wanted to bring a mine into production, and I just did it."

A lifetime achievement fulfilled and, all of his hard work – keeping the company afloat and out of debt, and in possession of a world-class moly deposit, throughout the darkest market days – will remain of constant benefit to Win-Eldrich and their shareholders for decades to come.

It's no wonder that Reuben Brant seems like such a content and optimistic man. "You have to be in good spirits – if you don't keep active, you'll grow tired."

And in closing, "If it makes you happy when you do something that helps the shareholder, and provides employment in the area, then I am," he says, knowing that his legacy is a successful, producing moly mine; an achievement to be proud of.